Brainstorming

Guide to Making Brainstorming Sessions Effective

(Generate Ideas and Solution Proposals for Daily Problems With Your Team)

Timothy Boyd

Published By **Simon Dough**

Timothy Boyd

All Rights Reserved

Brainstorming: Guide to Making Brainstorming Sessions Effective (Generate Ideas and Solution Proposals for Daily Problems With Your Team)

ISBN   978-1-77485-991-9

No part of this guidebook shall be reproduced in any form without permission in writing from the publisher except in the case of brief quotations embodied in critical articles or reviews.

Legal & Disclaimer

The information contained in this ebook is not designed to replace or take the place of any form of medicine or professional medical advice. The information in this ebook has been provided for educational & entertainment purposes only.

The information contained in this book has been compiled from sources deemed reliable, and it is accurate to the best of the Author's knowledge; however, the Author cannot guarantee its accuracy and validity and cannot be held liable for any errors or omissions. Changes are periodically made to this book. You must consult your doctor or get professional medical advice before using any of the suggested remedies, techniques, or information in this book.

Table Of Contents

Chapter 1: Inspiration And The Nature Of Inspiration

In our search for the most innovative ideas and innovative solutions it is important to first understand what the definition of inspiration really is. It is essential to know where it originates and know how it interacts with us. Simply put that, inspiration can be described as the experience of receiving an infusion of creativity that allows to generate exciting and fresh ideas. Inspiration is the basis of creativity. It is the source of the feeling of enthusiasm and readiness to take action. A person who lacks inspiration is an artist who does not possess the creativity needed to fulfill her ambitions. A businessperson who lacks vision will not view his business through an eye for creativity and may not be able to realize his potential as leader. We require inspiration to become creative people.

If we are dependent on inspiration to accomplish our tasks in a unique manner Where does inspiration come from? It is often believed that inspiration is a type of mythical creature, such as the unicorn who is hidden in the distance and we have to search for it meticulously. If we are too loud the unicorn will flee away from us, and we won't be able to find the inspiration we seek. Some believe that the best method to inspire yourself is to just sit and sit and wait for it to come. They believe that the mythical unicorn will come to visit them and, until they receive an appearance from their source of source of inspiration, they can't accomplish anything. However, we are in a bind if we were to behave that in this manner, as it makes us dependent on an external force to aid us in our tasks. The notion that inspiration comes from an external force, puts our in a state inability, dependent upon the dictates of something we can't manage.

You may not believe it that the view from the outside of inspiration is among the most well-known views there is. A lot of people imagine an artist sitting in a corner waiting for the right moment or time to strike. A lot of people believe that inspiration occurs from the middle of the night for the poet, artist, or even the person who is creative typically in some sort of a fantastical way like a dream coming to them or an uncanny image of the future. The idea of inspiration comes mostly from a romantic look at the process of creating. The artist is not visible to the public. working for hours hours working on their project continually revising and adjusting how his work appears. The public doesn't see the initial, second or third drafts by the writer, they only look at the final product.

If an idea's process is omitted the notion of someone being suddenly amazed by a huge and bold idea appears like it's something magical. However, this perception of inspiration blocks the ability of our own to generate ideas completely. This view from the outside is extremely popular however it is based on a important assumption. This assumption states that we are unable to influence or generate motivation, instead, we are in the hands of some external force that comes whenever it seems to. What do these false assumptions mean? Let's look them up.

Inspiration Assumption One: certain people can have it.

We frequently hear the expression "oh I'll never be innovative," or "that man's far more gifted than I will ever be." In many cases, there's an implicit belief about the ability to think creatively or with ingenuity. This is based on the false belief of it being inherent in a particular individual. This is usually based on the notion that external forces of inspiration is more than those who are creative. Instead of trying to refine and develop their own perception of creativity it is easy to and quickly attribute it as an inherent talent that the artist does not have control over. This can lead to think that they're incapable of being inspired or creative and thus they aren't inclined to investigate their own abilities.

However, if we examine those who are highly imaginative, or have huge ideas, we could see patterns and trends within the individuals. They usually spend a lot of time working on their projects. They tend to be very loose and honest in their work. They possess the spirit of fascination and curiosity within their own. They don't fear failure. If you were to look at every person you know who is creative and you could see that they appear to be frequently awed by their creativity and that's due to the fact that they're usually in locations in which they can discover an inspiration. An artist isn't any like any other but they're usually trying to reach their goals of being creative and, in the process they develop a greater sense of creativity.

Inspiring Assumption Number Two: The Creativity simply happens

The notion of someone exclaiming "Eureka!" as they suddenly come up with a great idea is deeply embedded in our modern notion of creativity. We tend to think of instances of inspiration as isolated events, which creates a greater mystery in the concept that being creative. The idea that comes up unexpectedly is not a thing that can be controlled or imposed. The idea of creativity as an uncontrollable lightning bolt that hits an individual makes the thought of developing a mindset that is creative to be impossible. How do we cultivate the ability to be creative if it was an event that occurs in a random manner?

The truth is that the spark of inspiration isn't an isolated event. It's an ongoing process that requires some time. The process of creativity leads to the single second when you say "aha!" but it requires time, effort and concentration to get there. Take a look at the way Thomas Edison invented the lightbulb. Edison didn't get up in the morning experiencing some sort of a bizarre dream that instructed him on how how to make a lightbulb rather , he spent thousands of time trying out different ideas. The process eventually brought him to an "aha!" moment where it was possible to discover how to make the lightbulb function.

It is evident all over the history. If there is a famous person who has discovered something truly amazing, there's an immense amount of effort behind the discovery. Inspiration at the end the day is a result that is constantly working. If you work to improve your creativity You will eventually discover an inspiration. A well-known myth regarding Isaac Newton was the way Newton discovered gravity. It is said that he was in the shade of the tree when an apple fell onto his head, prompting him to be aware of gravity's existence. Of course, things constantly falling around Newton's time So was it the first time that the man had seen something fall down? Not at all! The actual story lies in thinking about how gravity works initially, and when he witnessed the apple falling down then he began to think about why they fall perpendicular to the ground. The mind was already creating ideas, but the trigger (seeing that apple) was enough to cause the imagination to fire. That spark of light led to the igniting of all his ideas that he had in his mind.

The third assumption of inspiration is that it is an emotion

A risky notions that you could make in the world of creativity is the belief that you need to be in a particular kind of mood in order to be creative. If you're looking for a great idea , but you're not feeling particularly in the moment, you might delay your task until you feel the need to it later. This is a risky assumption because it's a result of something that can't be managed. It is impossible to control our moods and our emotions. We don't possess the power to trigger an emotion or mood at request. Therefore, when you are desperately in need of ideas but you're not in the mood, you could be in trouble.

The notion that inspiration comes from emotions or moods is incorrect. Inspiration isn't derived from an emotional source, since it transforms inspiration into an external force is beyond our manage. Inspiration is a result of the long hours of hard work and a constant cultivation. In the event that we hold off waiting until the appropriate mood to hit us, in order to complete our creative tasks then we'll never finish our work completed. Imagine the situation if a well-known and prolific writer was to be patient until he felt like writing. When will his work be finished? It is a fact that the majority of efficient and innovative work is completed no matter the state of mind of an person. When you hold off until you feel in a mood to write, you'll be in danger. If your writing is done when you are motivated, you'll encounter something totally different. You'll realize that you don't have to be enslaved to your feelings, rather you'll be able to complete more work and increase your sense of inspiration.

We've also discussed the most dangerous assumptions you make regarding the creation of an atmosphere of inspiration in your daily life. We've learned that inspiration isn't a purely external phenomenon that appears on its own , and disappears without the permission of the creator. Where is inspiration really coming from? It originates inside the person! The essence of inspiration is that it's an internal process and not an external occasion. It is a process that takes time and doesn't occur instantly. What exactly do I mean by saying that inspiration is an ongoing process? Let's think about the issue in this way instead of thinking that inspiration is an imaginary creature which you need to hunt to the ground, imagine that it's actually a plant. Every person has different kinds of plants, and all have different sizes, but we all have access the same tools that nourish the plant. Our attention to our own motivation and levels of creativity will determine how inventive we will be over the long term.

This is a great thing! This means that you've got the capacity to be as inventive as an outstanding artist, you'll be able to generate concepts like da Vinci and you'll are equipped with the tools needed to be the Steve Jobs type of industry and innovation! However, having the right tools available to you isn't enough to come up with brilliant ideas, you need to know the tools available and how you can use them. We discussed the way inspiration is more of the seed of a plant rather than unicorns So what are the tools are needed to nurture and cultivate the inspiration plant? Let's jump to Chapter 2 to learn more!

Chapter 2: Tools Of Inspiration Tools Of Inspiration

We've discovered that inspiration doesn't occur in one, single incident, it's time to take a look at the implications. The result is that we can increase and expand our own ideas. What can we do to create the mental state that allows people to become more imaginative within our daily lives? If the tools for creativity are already within us, how do we get them?

The first step is to begin looking to inspiration as something is already available to you. You're no different from anyone else in the world, even those we think of as truly amazing and successful. The world and the past both love to see the well-known and talented creatives as exceptional and unique, and frequently we get caught up watching someone else's life with the belief that we'll never be as successful or creative as theirs. The reality is that, even though we're all unique and unique but we are composed of the same elements. We all have different characteristics and personalities, however, we share the fact that we're all human. Everyone is within reach. Everything can be taught. You need to believe the fact that you are an innovative person and that you are able to come up with innovative ideas effortlessly.

If you've made the choice to view your own creative potential You'll then get an opportunity to gain knowledge of the techniques and tips that can help you inspire. Let's get started and take a look at the various tools that we have to draw inspiration from.

Inspiration Tool One: Curiosity

One of the most powerful instruments in coming up with innovative ideas is to ask the question of why. The word"why" is commonly used to prod and poke to help you gain knowledge concerning the universe around you. If a kid is young, he starts to ask questions about why and usually seeks to comprehend all that is happening around him.

He develops routine of interested about the world, and even if that curiosity is punished by adults, he'll be curious and interested. However, something changes during the process for the child who is curious. At some point, his parents become annoyed to hear him ask why constantly. It is possible that he won't get answers as he gets older, but then the question is promptly answered by "because I said that."

The interest in learning decreases after we enter the education industry. Once our education system has us into consideration, we are taught numerous things and information is provided to us. We cease to look for answers as teachers' job is to give information to us. Thus, our curiosity decreases. Then we begin to find many answers, and then cynicism begins to creep in. The cynical outlook of teenage years replaces the concern of what is wrong with a smug attitude. In the process, teenagers begin to revolt at their own parents. They ask questions about what they believe in and their parents usually are scolding them for asking questions. This can lead to a tense situation in which the child starts to feel pressured to be into line and stop asking questions. In adulthood, asking an unintentional question in the wrong place at the wrong time no matter how innocent it might be it can end a career. We learn to choose safety over of seeking out curiosity. Our curiosity diminishes as we gain knowledge about the world, and consequently we lose a vital part of our capacity to be invigorated.

The issue of why is one that makes us examine the fundamental assumptions. Curiosity permits you to ask questions without judgement. Take for instance Henry Ford, the creator of the model T. He was intrigued enough to inquire about why we were required to move by horseback. He inquired if there was an alternative way to do it. As the line of assembly was designed the process began with the question of why. What is the reason why we had to construct things one by one? The person who designed the assembly line came up with an innovative idea, but it was based on one question. What is the reason we operate this way? These questions stimulate curiosity, as well as the act of answering these questions will give you a an increased chance of discovering.

It's the process of finding an answer that provides the possibility of inspiration to develop. If you continually question what the reason behind things how they are, you're giving your mind the space to contemplate the assumptions that could be untrue. If a false assumption is found, and you discover that something could be improved and that's when the inspiration starts to rise.

If you'd like to remain in a continuous process of finding ways to improve your life and develop innovative ideas then you must be constantly in the state of fascination. Henry Ford once said "If I had asked individuals what they desired they would have demanded more speedy horse." This mindset that is known as that of "faster horses" approach, is commonplace in our society. People aren't looking at their world in a new way and aren't interested enough to know the reasons behind what the reason behind why things work as they are, and in the end their curiosity dwindles. They don't have inspiration, and they don't feel inspired since they don't have anything that they can be inspired by.

Inspiration Tool Two The Purpose

The most important thing to find innovation and inspiration is to come up with a compelling motive to be inspired. It may seem like a waste of time, but consider it for a moment. According to the saying, necessity is the root of innovation. If you're looking to inspire, there must be something in your life which requires the inspiring. An exceptional artist or inventor isn't one who creates simply because they believe they are legally required to. Some of the most remarkable creations that have ever been made and some of the most significant human achievements didn't occur due to a feeling of obligation however, they gained an increased sense of purpose from it.

Think about The Wright Brothers. They were not able to finance their venture or formal education, and nobody surrounded them while they worked on their amazing plan. Their plan? It was to construct an airplane and their most fervent desire was for it to become a the point of. This passion for the project energized the team and challenged them constantly and forced them to constantly look for inspiration to help to come up with new ideas.

If we are to develop an elevated sense of motivation, when we are looking to create amazing ideas, we need to be able to sense the need and the goal. There must be a drive that is deep in our souls. If we do not have this motivation and passion, we are at risk of just looking at our work with a low-level of approval. Inspiring lack of passion can make it impossible to get motivated because if you're not driven, there is no reason to do anything at all. Without a clear purpose there is no motivation to constantly improve. The pressure to be better and strive to be better and fulfill one's goals is the key to creating more motivation. Imagine this in terms of The more motivation and sense of purpose that you feel and the more you're likely to be able to accomplish your objectives. Innovation and creativity are stimulated by a heightened sense of concern about the work you're doing and if you're not located in a space that you feel an underlying direction, then most likely won't be able to come up with ideas.

This is the reason it's so difficult to incite this kind of creativity in the work environment. The majority of people aren't satisfied at their job and don't believe there's any meaning behind it. And with no sense that we have a purpose to live, creative ability decreases. That's why successful businesses work hard to give their employees a the sense of purpose so that they are better able to be innovative.

If you're experiencing that you're not very innovative in your chosen area, or have trouble coming up with concepts Consider whether you truly feel that you are motivated by the work you're doing. If you're in a situation where you're not feeling motivated because you view the job as boring or insignificant, you're likely to not find enthusiasm until you are able to alter your perception. Sometimes, it's the matter of understanding how to change the way you think.

The Inspiration Tool Three Perspective

Two men were who were laying bricks on the other roadway's side. When asked what they were working on, one declared "I'm building bricks" while the other declared "I'm building hospitals." The two men were both doing the same thing, however they had completely different perspectives. These perspectives influenced the way that men worked. One person is willing to be extremely dedicated due to the fact that he sees that his work is extraordinary, while another person might be working extremely hard, but not be as motivated and inspired than the person who is focused on the final objective. If you're looking to build an unwavering sense of innovation and creativity it is essential to acquire the mindset to see the work you're doing from more thought.

This is what we call "high high altitude thought." Imagine what the world will appear after you zoom out and take a look at everything around you. Think about how a football game appears from the perspective of a player and a spectator who is in the spectators. A player's eyes are only the area in his immediate vicinity, therefore his capacity to think of strategies is limited by his own perspective. He only sees the area to the left and right of him but not what's behind and not even what's in front of him. However, a person standing in the stands is able to see the whole game. He is able to see the game from a greater perspective and, as such, it is possible for him to make plans regarding the entire scene. This is the reason why the majority of football teams have players who can observe the whole game from a greater view and provide advice to what they're experiencing.

Have we got this type of attitude when we are thinking about the issues? Sometimes we get distracted by what's before us that we fail to think about the larger perspective. But perspective is what allows us to build the mindset of a big picture. A willingness to step back and consider every moving part in the search for a solution can boost your creativity considerably. If you're hoping to enjoy unending innovation, you must learn to expand your perception. If our perspective is limited and we don't have the capacity to capable of coming up with innovative ideas or unleash our imagination. Perspective is what can define the individual.

What exactly do you mean by having the ability to transcend your perception limits? The answer lies in considering what a problem is seen from a different perspective. For instance, if you're trying to find a way to reduce the cost of shipping an item for your business It is possible that you will be interested in reducing costs of shipping costs by negotiating the lowest cost. You shop and search around but you can't get the product you're looking for since it does not exist. One way to get an understanding of your situation is to take a look and take a look at the bigger view. It's not about getting an affordable shipping service but to have the lowest shipping costs and, therefore, you may require changing the material to lower the cost. Maybe you'll need to determine how to move your distribution centre to an area that's simpler to move your shipment from.

Perspective helps us avoid our natural tendency to be obsessed with a particular aspect. It is often said that "he can only observe the trees but however, not necessarily the trees." The focus on what is that is in front of you can hinder your ability to perceive the bigger view. It is important to take a look back, to consider every aspect, to take a look at every part of the process and to ask yourself a series of questions concerning assumptions will help you see the bigger picture of the world around you. The more willing you are to let go of focusing on only one aspect and attempt to change the way you think about your issue, the greater likelihood you'll be able to achieve having a breakthrough.

There are a myriad of methods to change the way you think about a problem. For instance, you can sit down and write down what's happening and make a list of all possible options. Try to come up with solutions that don't appear on your list. You can bounce your ideas off a buddy who is willing to question your ideas and offer a different an alternative perspective. You can even stop what you're doing and attempt to do something totally new for a few minutes Then come back to take it in with new eyes. A part of learning to see things from a different perspectives is to change things up so that you don't stay to your usual thoughts. The typical human pattern of thought is to be too focused on what one already knows. The goal is to gain knowledge about something you aren't aware of. This is why you must constantly be on the verge of shifting your perspective. This will increase your creativity and inspiration in the course of time, which you would not imagine!

Obsession: Inspiration Tool Four

If you're hoping to be inspired by something, then you must to be enthralled by it. If you take a look at any of the famous individuals in the world with an unending supply of new ideas, innovations and ingenuity, they possess a strong sense of obsessiveness about what they are doing. Obsession extends beyond the realm of interest. There is a lot of jokes about an item being an obsession, or look at someone who is obsessed with something and claiming that they are obsessed however, what exactly is it that means being obsessed?

A healthy and true obsession occurs when you can't put off thinking, reading or learning about the subject. In this type of obsession is that you'll find the most creative sparks. The more you study and digest and the better you comprehend the subject at hand the more possibilities are envisioned. If you're someone who hasn't committed himself to learning all you can about a subject and aren't looking to gather the most information you can on the subject you're looking to develop thoughts about it, then you're seriously restricting your ability to think of new ideas.

If you are able to think about, discuss and obsess about something, the more chances of gaining an inspiration. An excellent designer doesn't have to be because he was created this way, but he's inventive because he's never stopped thinking about the process of design. Highly skilled athletes and performers have the qualities they do because they are trained with an obsessional mindset. It is often emphasized that we should not getting too involved in things, and we are advised to keep the balance of our lives. However, if we're hoping to be successful and realize our goals through innovation, then we'll have to face the fact that an average level of interest in something isn't going to produce exceptional outcomes. It is essential to be obsessed with what we're doing. Do as much reading as you can, speak about it with other similar people, and find ways to keep contemplating, planning and researching. The more you consider your project and the more your brain will become habitual of contemplating your task. This will be transferred to your subconscious mind, and, over time, you'll notice that you are able to quickly and easily discover ideas and inspiration on the subject.

Consider obsession in the same way as water. As much water as you allow your brain to think about, the more room for swimming. If you don't have plenty of pool water, you won't truly go swimming anyplace. However, if your pool is completely submerged and full of water you'll be perfectly. The desire to be immersed in your thoughts throughout the day, and it will take over you until you're in a perpetual situation in "swimming." That will help you develop an appreciation for inspiration in the course of time.

Inspiration Tool Five The Competition

Humans tend to be competitive animals. This can stimulate our minds and provide us with an energy boost when we set out to achieve the target. Competition can also spur the development of new ideas. In the time that Henry Ford invented the Model T He wasn't looking to change the color of the vehicle from black to any other color. Ford didn't care about what people wanted and chose to stick with what he liked. What happened when other companies began making their own vehicles with different hues? This forced Henry Ford to release a new model T in the same color. In the course of history, you will observe how competition has dramatically transformed or improved the design of a product. Take a look at the fight with Apple as well as Microsoft. Both companies can't afford to be lazy, because when they did, competitors will catch up to them. This makes them always in a state of ingenuity. This is an excellent creativity that makes it possible people to transcend their own self-confidence.

If you're looking to build a stronger sense of motivation and be more creative, you may be interested in adding a competition to your daily life. Maybe if you took an unintentional bet with a friend or even challenged one to a contest or contest, you may be able to experience more enthusiasm and excitement. Nothing will stimulate an individual more than a competitive spirit. If you're in a real-world field in which you must be competitive and you are in a position to be competitive, you may think about learning more about your competition to show you how to perform superior to them. This will increase your capacity to think about things in a new way and give you an advantage over your competitors.

Sometimes , we may dismiss the idea of competition simply because it's not pleasant to be competing. But it's the sense of competition that makes it possible us to attain the highest levels of success. Even if you're in competition with another person does not mean you should dislike them fight with them, or be negative toward them. It's simply that you're trying to be better than them. They'll strive to surpass you too. It's part of human nature, and you could be tempted to profit from it!

Sixth Inspiration Tool: Momentum

Sometimes, one of the most powerful instruments in the arsenal of inspiration could be simply the power of the momentum. There are many occasions where someone is waiting for lightning to strike them. They are eager to get a bigger amount of interest by the unicorn that inspires them however, it doesn't come.

They shuffle around , hoping for something that isn't going to be delivered. Don't do that! Try to think about how inspiration grows significantly in those who have the most momentum within their daily lives! What is momentum? It's the process of moving forwards as you work, even when you're not feeling very inspired.

Inspiration and creativity occur with the course of. The longer you put in at it, the more chances are you'll allow your energy to propel you to an inspiring moment. Instead of awaiting something that isn't going to happen decide to focus on your goals and remain consistent in your endeavors. It may take some time, and it could take longer than you'd like however, eventually you will get to a point where you are inspired. You'll have that brilliant idea, and you'll be elated and filled with imagination. If you're not constantly striving to achieve this goal, if you aren't coming up with fresh ideas each day, there's not much of a chance to get that type of motivation. You must be continuously working towards your goals to be in a position that inspires you.

That means you shouldn't depend on your emotions or moods to motivate you to go to work. It is essential to commit to working every day, even when you don't feel like it. Stephen King believes that writing regularly will be more effective than writing only when you are motivated because if you wait until you feel feeling inspired, you'll never achieve your goals. If you only feel inspired every week, you'll have to endure six days of waiting. If you are working for six days, and on the seventh day, you come up with this amazing idea, you have six work days ahead of you. It's that easy! So, don't sit around waiting for a moment or a moment to hope that Lightning strike, but instead create momentum by working each day. The more consistent you are at it you will come up with more ideas and new ideas you'll come up with.

7 Inspirational Tools: People who inspire you

If you're looking to be more motivated in your daily life, or to develop a higher level of excitement , so that you are able to unleash your creative side, then you might think about seeking out inspirational individuals. There are plenty of inspirational speakers around the world. It's not very hard to find some great podcasts on motivation and inspiration. There are many books that are devoted to this topic. If you're not using these tools, you're wasting an essential source.

Many times, you'll encounter the same scenario When someone listens to an inspiring speaker, they are enthusiastic and on Monday the same thing doesn't happen. They die out. It is possible to consider someone to be an inspiration addict, rushing between speakers, one book to another, and method of method to method and never altering their lives. We aren't looking to be that way we do not want to draw our entire sources of motivation from outside items, instead we wish to enrich our lives with these ideas that inspire us. Supplements can be helpful but aren't the primary source of our nutrition. Keep in mind that we are trying to plant a seed here, and we're trying to encourage our motivation and creativity grow, which is why we should sprinkle plenty of water over it. So, don't consider inspiring speakers as the primary source of motivation, instead consider them as a tool that can assist you in your quest for motivation.

It's great to stay in a constant mode of consumption of the things that make you feel inspired directly. A constant intake of speakers, podcasts blogs, book reviews, and other blog posts keeps you on the move even when you're lacking inspiration. Sometimes, they can give you the right boost to allow you to break through an obstacle. It happens to us all but there are times that are really difficult to push yourself. Therefore, providing yourself with tools to aid in boosting your motivation is of paramount importance.

Another way to get inspiration from the other people. If you meet people who are similar to you or is with the same interests that you're keen on, and you're with people who inspire each other and inspiring, then you will be able to profit from this. There is a lot of value in being a part of and collaborating with those who are inspiring to you. They can assist you to achieve great results and in return, you'll be capable of helping them in turn. If two people are in a group and they are able to have a greater possibility of pushing each other further as opposed to if they were on their own.

Humans are social creatures , and as such when we start working with other people or groups it is possible that when the group is healthy and positive one, our imagination and creativity can skyrocket. This is great because it allows us to surpass our inherent limitations and perspective. Nobody is alike, and they can be challenged and push each other towards greater heights. This is a wonderful aspect to be able to experience, and will create the ideal atmosphere to foster innovation. When you look back through the past, you will see numerous people who had great colleagues and friends who supported in their endeavors. If you partner with an accomplice in crime, you increase your capability by two times. This will be helpful when it comes to generating new ideas!

These are amazing tools that can aid you in developing the ability to be creative within your daily life. If you'd like to increase your creativity and discover ways to think outside limitations, then these tools are the ways to think that can improve your sense of inspiration you. Keep in mind that inspiration must come from within, and it's an ongoing process that is not a singular occasion. Therefore, if the process of developing great ideas does not happen instantly, but is rather an ongoing process then what else do you have to think about? We've spoken lots about the many good things that can help you nourish your creativity and inspire you however that's only a small portion of the story. If you help plants in their growth it, don't simply put water on the plant and then let it dry in the sunlight. Make sure you get rid of decaying branches as well as keep insects away from harming the plant. If we wish to maintain strong, healthy levels of inspiration We must be prepared to examine certain factors that can harm our motivation. Let's continue to the next chapter, and look at the factors that can kill our motivation.

Chapter 3: Inspiration Killers

Inspiration is something that must be nurtured inside however, there are plants that could sprout and choke our creativity. They choke and destroy our creative juices and before we even begin to think about your ideas or dreams there is a chance that we will experience an absence of motivation. Therefore, to gain a better feeling of creative and inspired, you should be prepared to consider certain factors that undermine our creativity. These issues, whether we like they may, will always originate within us. There are external events that happen to us however, at the end of the day, it is us the ones that decide whether we allow them to impact us. Let's take a look and begin to look at the numerous ways to kill inspiration.

The Inspiration Killer The reason is lack of confidence.

Have you ever had an amazing idea, but realized that it was unattainable? Have you ever found yourself in a position where people have undermined themselves simply because they don't believe they have the capability of executing an idea? This is due to the lack of confidence in your own abilities.

If you're confident, you're giving yourself the right to think of the many ideas you want. If they come up and seem feasible, your confidence is what makes them emerge. For instance, if you had an idea of how you can save money in your monthly cable bills If you believed in it, you could call the number and contact the company right away to take care of it. However, if you had an absence of confidence then you may attempt to convince yourself that you're not going to succeed and not even try it! As time passes, and you go about your life with any sense of confidence, you may notice that your thoughts aren't coming into your the forefront as easily or quickly or easily. Why? because you kill them right immediately due to the lack of faith! The less confident you are, the lower of a chance you'll come up with a fantastic idea. Instead , you'll believe that the idea you come up with won't be successful. It's a perception issue completely.

A person with confidence can rationally and accurately evaluate his ideas according to their merits. He could choose to rethink a course of decision based on what is known to be the truth however, it's not arising from lack of confidence. It's more likely to come from the security of knowing that the thought is either positive or negative. However, someone with no confidence isn't going to possess the ability to distinguish between good and bad, since they don't trust in their own abilities, and so consider everything they see as bad. As their lives progress they fall into an emotional void in which they are unable to come up with good concepts. They blame themselves for their failure to think of new ideas and their confidence is waning.

This vicious cycle is an everyday reality for many people's lives. There are number of people of us who are told their ideas aren't good enough and, as a result of this, get the feeling that they're not doing enough. Therefore, they quit trying at an unconscious level. In reality, they're still trying to think of new concepts, but the core part of themselves has abandoned the idea. This causes a feeling of suffocation that a lot of people around the world get stuck in.

The answer to this is to master the art of developing confidence in yourself. The best way to do this is to learn to be able to distance yourself from your thoughts and not criticize them with harshness. Instead of dismissing something in a hurry, consider looking at the issue as if a person was asking your opinion. We alter the way we view things when others ask us to share our thoughts. Self-judgment can be a risky foe when you're trying to come up with fresh ideas, and If we don't feel confident about ourselves, then we may judge too severely.

Self-confidence is a process that takes time however, it is based on you making the decision to be kind to yourself, and stopping constantly rebutting your thoughts. Instead of immediately denying the things you say or think about, you can run through it with another person. Write down the advantages and disadvantages. Do not let your ideas go to waste when you've got them, it will result in ruin. I'd wager that the majority of people who believe they cannot come up with good concepts actually come up with plenty of great ideas, but they lack the confidence to pursue their thoughts. Avoid falling into the trap of dismissing your ideas prior to trying your ideas. Test the universe your ideas, but don't kill your own ideas!

The Inspiration Killer 2 Fear of Failure

Failure is among the most misunderstood concepts in our world today. There are lots of people with an inherent fear of failure due to the fact that they believe it's the worst thing that could occur to them. This fear can hold them back and stifles any creativity or ingenuity that a person could be able to. Failure for many people appears to be more terrifying than any other thing. If you are afraid of failure, then how do you let yourself be free to learn from your mistakes?

In all honesty there's plenty of great ideas out there. If you're struggling to find a way to put something into practice or you're seeking more creativity and inspiration, the process of generating great ideas isn't necessarily the most difficult part of being an innovator. The toughest part of being innovative is the implementation. The art in implementing ideas and making your plans a reality can cause your stomach to tighten up in the fear. Most people find that failure is more damaging than doing nothing. This is exactly the situation! People are scared to death that they're not able to commit to something , and then complain about their lives for not having the plans they've been longing for.

If your fear of failing is subconsciously rooted and you're afraid of failure, it's even more. You may do all you can think of innovative concepts, but your own anxiety will permeate each one. You may not believe that you've got a great idea because you're afraid to even try to implement these ideas.

What is the source of fear about failure? originates? It is usually rooted in living in a society where failure is treated as a punishment rather than praised. If a child fails on their report card, the child is scolded and criticized instead of being shown the importance of learning from the mistakes of his past. If a businessperson fails in his job, he's fired or threatened rather than exposed for his errors. In our society, that is where people are praised for their achievements, however every failure is swiftly blamed. It is then a shock to discover that we find ourselves in a society of individuals who have been scared to experiment with something new. What is true is that failures can result in massive achievement in our lives as we can take lessons from these experiences. However, if our view of failure is always one of avoidance and fear is denying an essential element that is human.

As human beings, we will always blow it all the time. There are times when we're bound ruin everything and at other times, the effort fails completely. If we change our view to see these events not as a negative or scary, but consider them as positive and essential events that happen within our daily lives we're putting ourselves in a position of growing from these experiences. Every mistake can be an opportunity to gain knowledge from the mistakes of others. If you make a mistake then you're able to sit backand review the circumstances and ask yourself what caused it? There is a great chance to gain knowledge from your mistakes and discover ways to avoid mistakes at a later time. This is an excellent tool.

The most successful people on the planet have prepared themselves to fail numerous times. They've failed repeatedly but they didn't give up. They didn't give up, and because they didn't let go of what it is which is the most important thing for them, their drive to move forward was never diminished. they developed because of their mistakes. In a world that values success more than failure, we've gone wrong. Success is only a result of failure. The definition of success isn't a lack of failure, but an adequate number of failed attempts that result in expansion. There aren't many great individuals in history that haven't failed. However, we rarely hear about their mistakes. We hear only about their brave deeds and stunning actions. Politicians conceal their mistakes from us, and parents refuse to admit that they have failed, and even the slightest hint of failure creates fear in us.

If you're looking to be truly creative If you are looking to come up with amazing ideas that can revolutionize the world, you have to be prepared to be a failure. Not only should you be willing, but you should be anticipating it. If you are afraid of failure that you are afraid of, then you'll never take on the challenge of success. In order to be successful, you need to fail. That's acceptable! Avoid believing you're failing is the only bad possibility as it isn't. The worst thing that could be happening is that you make decisions based on fear. one of the worst things that could occur is that you run away from your brilliant ideas because you fear falling over completely. Allow yourself to mess up, and then make a mistake! You'll gain more knowledge from a thousand mistakes than from one successful experience.

Inspiration Killer Three: Daydreaming

It is a great way to get your mind off the task of in generating brilliant concepts and coming up with innovative strategies and plans. However, there is risk when you rely only on daydreams to come up with concepts. As mentioned earlier there's no lack of brilliant ideas around and you've probably experienced plenty of brilliant ideas. The issue arises when someone becomes focused on their ideas and not trying to put them into action.

There's a certain amount of daydreaming needed to come up with an idea that is innovative however, it doesn't end there. One of the biggest factors that kills ideas is letting it sit for too long in your mind. It will eventually begin to decay in the background. Additionally, daydreaming and fantasies could actually hinder your efforts in implementing your plan. Why? Because the human brain isn't able to distinguish the real from the imaginary. If you imagine how wonderful it would be to finish the project and complete something you've begun, your brain begins to think that it has accomplished it. It gives itself dopamine. It's a chemical that makes you feel happy, even though you're not even getting started. The reward system you are rewarded for which you didn't actually achieve could hurt you over time due to the feeling of achievement that result of daydreaming could be able to notice that your motivation has drastically decreased. It can take a long time to kill your motivation and ambition because you're always in the state of thinking that you've already accomplished something.

If you're having trouble finding a way to inspire yourself to do your work, however, you have no issues in thinking about and daydreaming about your next step This might be the reason there is no motivation. Your brain is wired to think is that it has already decided that it has accomplished the things you're fantasizing about and therefore has no motive to offer you any additional motivation.

If that's your issue, what do you do? Work! Really! The time to fantasize and daydreaming can be beneficial during the initial stages of a project, however If you don't start working immediately and start working, you'll end up destroying the dreams you've been imagining in your mind. The idea of inspiration has a expiration date, and if it sits in your mind for too long, it could never be seen in the daylight. If you've got an excellent idea, or whenever you feel the spark of you are inspired or you are looking to come up with a innovative way to accomplish something, get started immediately. This will shift the messages in your brain to stop worrying about the future and help you concentrate on the present, which is where the actual work has to be done!

Inspiration Killer Four: Negative

People

There's nothing more annoying than someone trying to ruin your parade. The sad reality of life is we are living in a society that is filled with selfish, negative individuals and some of them might even be trying to undermine your plans and dreams. They could mock you and ridicule you. They might even tell you your idea, and you'll become the very first to smack it down or tell you the reason why it's not working. They're not coming from a place of support or sympathy neither. There's a lot of benefit in talking with anyone about the issues you encounter in their work, or constructively criticizing something, but this isn't what I'm talking about. I'm talking about those that are extremely harsh due to the fact that they have a chip on their shoulders. They can harm your goals and could cause harm to your ideas if aren't cautious.

What makes these people so negative? It's typically a result of people feeling bitter about their own shortcomings and anger over something they feel that other people ought to be able to accomplish. Their thinking is "well when I'm not capable be able to, then why should anyone else!" This toxic mindset is a habit that can develop within an individual over time, and eventually result in the person completely ignoring the dreams of others. As you attempt to express your thoughts with enthusiasm, they'll go on to get all the enthusiasm and excitement from you.

People who are negative should be eliminated completely from the lives of you as quickly as it is possible. There is no obligation to stay with these types of people, and there's no law that requires you to discuss your goals or dreams with them. You can choose to leave relationships that are harming you and depleting your motivation. Avoid falling into the false impression that because you have been acquaintances for a specific number of time or they are family members that you must accept them. Obligation is not the best thing to do here. They use different kinds of obligations to make you accept their destructive and painful actions. Do not fall victim to those who have nothing else to do but to thwart other people's dreams. You're much more than what they are! Your work is important in the same way as you and if those who have a negative view of you don't be able to see that you are valuable, then you're better off making the choice to leave. They may be envious of them for it however it's better to let someone else be in their own devastation than be dragged down with them , too.

Inspiration Killer Five: boredom

There will be a point that you feel you're not really attracted by the job you're working on. You may feel it's a waste of time, or you may be bored, or be tempted to quit what you're doing and opting to go to a different job. This is among the most potent killers of inspiration that exist! If you are bored with your work , you'll not be looking at everything with interest, and you'll give up trying to see things from a different perspective. Instead, you'll be watching your watch instead. You could get stuck working on the task in a haphazard manner, and the result isn't as good as you would like. Your ideas may disappear since you're not interested anymore.

Boredom is a danger to creative thinking. If you're bored by your work, you're not going to have the enthusiasm, focus or energy required to move things forward. If you're stuck in a place that you are bored, what do you have to do? If it's a task you're usually quite excited about, you may get burned out a tiny bit. It's the best thing to stop your work and take a time off. You could focus your attention on something completely different or relax for a couple of days. It is a serious threat to the progress you've created in cultivating that sense of motivation Don't think of it as a minor thing. If you realize that you're extremely bored by your work it's time to take action.

If you're not able to take a break as an option, then you may think about finding something to make you not be bored by what you're working on. Sometimes, the feeling of being bored is actually an attempt to protect your body from discomfort. If you are feeling bored, you could be feeling concerned, anxious or worried and yet the boredom is hiding these emotions. This happens often, particularly when working in an office. Think about why you feeling bored, what can bring it to life and what can you do to overcome this obstacle. It is your responsibility to ensure that your work is enjoyable If you're in a place where you're not enjoying your work and you're not happy, then you must be the one to determine what you can do to make it better. Avoid falling to the temptation of becoming bored of your job and using it as an excuse to not work. You are the one responsible for your personal level of interest therefore, you need to find ways to make it enjoyable. You can make it an event, you can shift your work location and you can concentrate on a particular section of your job you like the most or you can be able to spend time imagining your success to motivate yourself to be more productive. It's not possible to choose to leave because you're not in the mood!

The Inspiration Killer 6: Irregularity

Inspiring people are the most dedicated working. Anyone who is consistently putting in hours after hours each day every month, will be able to find more energy than those who don't take the time to do anything all day. Therefore, the reality is, if you're not constant in pursuing ideas, and if you don't make time in your schedule to consistently try out the latest ideas and ideas, you'll never stand any chance. Do not let your schedule get overwhelmed with stress to the point that it hinders you from getting things done. Make it a priority to consistently work on your projects or your idea will never truly appear.

Jerry Seinfeld talks about a method of productivity called"The "Don't end the chain" system. It basically means that every day was writing, his calendar, he would mark it with a large red"X. He would then count the number of days they was able to write on a single row, and make sure not to cut off the line of X's were written onto his schedule. If you would implement a method like that and to not break the chain, even if it's just only a tiny bit every day, you will be amazed at the amount of energy to be found within your own life! Do not break the chain rather, work each every day to accomplish what you wish to accomplish! You will soon realize that you are more motivated than if you only do it occasionally. The brain loves routines and patterns, the more time you put into in a routine, the faster your brain will develop the habit of seeking motivation.

Inspiration Killer Seven: Stress

It can also be a fatal factor for our bodies! Stress over a long period of time and it can have an adverse influence on our sleeping patterns and mental health as well as your blood pressure! Stress can also derail your creative and inspiration If you're not careful to control it! Creativity requires some playfulness and relaxation to experiment. If you're overwhelmed and stressed, the creative pathways inside of you can shut down , and your thought patterns may become overwhelmed and negative. It makes being creative difficult when you are overwhelmed by all the negative events within your life.

If you're looking to become more innovative and think of great ideas then you'll have to manage your stress levels. The more stressed you are, the more difficult it will be to get work accomplished. If you've never made the effort to understand ways to lessen stress within your daily life, then you ought to. Not only is it good to your health, but it's very healthy for your spirit when you make a an effort to reduce your stress levels.

How can we lower your stress? It's up to you, however there are a couple of ways you can take to lessen the stress you experience that you experience in your daily life. Meditation is an excellent method of learning to relax more. The breathing exercises, the ability to clear your mind, and to sit in silence will help you discover how to manage your emotions and decrease the stress level that you experience. Another effective way to lessen stress is to exercising. The more you exercise your body, and the more you train and your body's body will release chemicals that help reduce stress. This will reduce the stress levels of your body and give you an increased amount of energy throughout your day.

Another great method to manage pressure is taking a few minutes to relax, go away to a getaway. If you're not in a situation to relax Try finding something you love and then getting involved in it. There is no need to always be at your best are. Sometimes, you can have a break and take a break and relax. Stress can be pushed upon us all by ourselves. Sometimes, we are able to take on too much to do or create an unneeded self-consciousness. It's good and healthy for your motivation level to take a slow down and breathe deeply and rest. Do not think of resting as lazy, but think of resting as an opportunity to prepare your body for the future.

That's all the big hitters when is about the things which can end our enthusiasm. If you are able to recognize and overcome these kinds of situations and overcome them, you'll be in a far better position to overcome your own challenges and become the source of inspiration you've always wished to be. Before we conclude this book, let's examine every step that can help you come up with brilliant ideas!

Chapter 4: Idea Generation Process

If you're hoping to create amazing ideas, or would like to make use of all your inspiration that you've spent time building up and grow, then you must be able to follow a procedure! There are many different ways to go about it in the creation of ideas therefore don't think that you're limited to the ideas we're going to show. Every person is unique and is using a different approach in the process of generating and implementing these concepts, so if need to add or change steps, it's at ease! We'll go ahead and outline all the steps in the creation of ideas! Ready?

Step One Step One: Ask Questions

The first step to coming to a new and exciting idea is to begin asking yourself questions. In order to think of some new ideas, you're searching for solutions, or are looking to make a difference in your surroundings, then you'll need to begin by asking questions at the simplest level. Consider as many options possible. Make a list of questions. If you are planning to create an innovative product, you should ask what people require. If you're planning to write a book, you must ask yourself, how you can do something different from other books that is out there? The process of creating a unique idea begins with the concept of an inquiry. The more questions you are able to come up with the more interesting! Do not focus on answering these questions just for now, ask these questions! Write them out and then look at the answers you've got! A variety of questions you could ask are:

Why is it this way?

Is there an alternative method to accomplish this?

What is it that makes this intriguing to others?

Who was the person who came up with this strategy?

How can I do this? to accomplish this?

- What's easiest?

- What's faster?

- What is the reason for it to be there?

Is it really necessary?

These questions can help you to find out quickly regarding the subject. If you're trying to design products or meet the problem, you're seeking to solve a question. The majority of people, when they purchase something, will ask themselves "why should I buy this?" It's your job to answer the problem, however it begins by asking that question!

Step Two Step Two: Answer the questions

After you've adequately asked the questions, and written them out You must then respond to the questions you've written down. It is possible that you will find that there aren't enough responses, however that's actually the greatest part! If you aren't able to clarify the reason or method, then you're probably looking for a different method to go about it! That's where the process of generating an innovative idea originates! Ideas that are innovative can solve problems. If you can solve a problem, you're doing an excellent job! Discover as many problems as you can, and then make a list of them. The process of answering is about identifying the specific kind of issue. For instance, if you can't answer the question "why do people frequently get their keys lost?" then you will be required to investigate the issue. This will open the possibility of a step 3!

Step Three: Study the Issues

When you've identified some patterns of thinking that aren't able to quickly be resolved then you'll need to conduct some investigation to gain a deeper comprehension of it. For instance, if want to know the reason why people tend to have lost their keys you could look up a few studies to discover that the majority key-lossers are due to the fact that they're not paying attention. This study gives you the chance for you, if you'd like it, to consider ways to resolve the issue. Keep in mind that the purpose of ingenuity and generating ideas is learning how to solve problems therefore, if you come across the problem and have studied it, you're required to think of solutions to the issue.

Fourth Step: Brainstorm possible Solutions

This is a larger undertaking than the other ones, but it's basically the exploration of ways to tackle the problem. The ideas will begin out of your head and you'll be required to effectively navigate through every thought. Keep in mind that there is plenty of concepts when it comes to the human brain, but the most difficult thing is figuring out how to sort through them all to find the most suitable one. When you start to figure out the possible solutions, when you think about your ideas, record all of your ideas that pop into your mind. Consider long and hard over all the possible solutions to the issue. As you think of new ideas, you'll realize that some of them aren't feasible and others aren't feasible however don't fret about it yet. Simply write them down. Try to record as many thoughts as you can. Do not fret about what is possible or can't be done!

Step Five: Sort Through the Solutions

After you've collected a huge number of options and ideas, it's time to make the decision to go through them all. There will be some solutions that seem feasible initially but will turn out to be unwise in the future. There are ideas that don't work, and there will be some ideas that will perform well after some tweaking. The goal of the sifting process is to make a judgment on the various ideas as well as their strengths. In other words, if you're considering six possible solutions you've come up with, you'll be able to assess how viable all of these ideas are. There are some that might not be, but it's the right time to separate them and concentrate on the other issues. If you've done the job correctly you'll be in a position to identify which strategies are actually effective and which will not. You'll need to move on onto the next stage.

Step Six: Select One

A person who is obsessed with everything isn't going to gain anything. Do not try to combine all the solutions in one huge package, rather choose one solution and then create the solution you want to. This is perhaps the most difficult part of figuring out solutions to problems since we tend to want to incorporate everything possible into one process, but it's not the most effective approach. The best approach is to concentrate on just an issue or problem at go. If you choose your one solution, you will have an increased chance of understanding what it takes to work it out , and create it rather than picking three ideas and tried to tackle them all simultaneously. This is likely to be one of the toughest aspects of finding a good idea, because when presented with many good ideas it is difficult to narrow down the options to the best one. You might think that in choosing one we're calling alternative solutions to be poor suggestions. Not at all! There are some great solutions however there's many good solutions in the world! There ishowever the issue of implementation therefore, if you are trying to decide between three or two options choose one. Making it happen is more important than having more ideas!

Step Seven: Experiment

Once you've come up with the solution, you'll be required to devote a significant amount of effort to work on the idea until it's actualized! Whatever your solution it's the time that you start to concentrate on getting it out of your mind into. If you're trying to develop an idea, it will require you creating the prototype. If it's a novel or play likely be a matter of writing the script. Implementation is a crucial element of being an innovator The idea you have developed is only one aspect that equation. It is essential to pursue it and also implement it! When you work on your new idea, you may discover the need for more thoughts that can help the idea, and that's great! It is advisable to allow your plans and ideas to expand as you progress. Do not lose focus on the things you plan to do however, as you go about your work take the opportunity to add to your current work.

The most important thing it is the flexibility and a willingness to try and experiment in what you're doing. There are times when things don't succeed however that's fine. The more you try and develop your idea and the more you're in a position to create the concept, whatever it's called it is, the more concrete your concept will become. If your idea comes out as a disaster it's okay! You have other options that you've created and all you have to do is revisit to work on a new solution. It is crucial to experiment with this method because it makes it easier you to determine the things that work and what isn't working.

Step Eight: Finalization

The final step is the finalization. This is when you've completed all planning, testing and testing phases. This is when you introduce your idea to the world to allow people to engage with. If you're an artist this could be the time you present your work. If you're trying to come up with a fresh plan for your business, this is the time to put into practice the plan. This is the final stage of the process. Thoughts start out as just ideas, tiny seeds, but over time, those seeds develop enough to be capable of interacting with the world. The entire point of having a great idea isn't just to be happy about your idea the main purpose of an idea is to communicate it to the world in a real concrete shape. The process of finalization is what makes it all come together.

Chapter 5: Forms, Paperwork, Applications, And More

"You're not stressing because you're doing too many things. Stress is caused by the fact that you're doing too many things that you do not see the value in."

There is a general consensus that the customer always is right. It is not true since the human being is a human and therefore not perfect. An alternative approach is to consider what we can do to focus on the client or maybe the user. What can we do to put the other first, in order to make their lives more convenient? Paperwork, no matter what form isn't a method to accomplish it.

Paperwork can be a hindrance to many things because of the emotional baggage that comes with it. We will look at various reasons why the endless forms we complete are a burden that affects your bottom line directly because of the loss of time and morale.

Apart from a small portion that makes up the majority of people, most people fear filling out forms, papers application forms, expense reports surveys, forms, or any other ordinary of this kind. An excellent example is your doctor's appointment. Have you ever had fun writing out some of the data, and then repeating it every year? It's even more frustrating when you discover that someone has already typed the data into a computer and simply want to find the differences. This makes it appear like a waste of effort that isn't it.

Many people know that there are legal as well as HIPPA obligations and the limitations of sharing network information in the medical industry which makes paperwork mandatory. However, some medical institutions have changed their practices to allow you to complete it only at a single time and then screen for yourself during your next visit for things which require updating. But, it's still each individual facility which you are required to take these steps. This is because the majority of people understand the idea that this kind of document can be a hassle to complete but in the end, it is beneficial to the person who fills it out.

The majority of the information else that is being completed is for the benefit of the requester and not for the person who is working to fill it out. It's not a process that is focused on the end user the reason is enough to make it a challenging task. The person who is doing the work does not see the value which is sufficient to be worth the effort. Lack of value perception could result in those who only provide the most basic information and not providing the information that really is required. It's an additional time-consuming things to complete, like taking surveys. Time begins to disappear rapidly when we type in information wrong, then must find the right data, leading to delays in processes that require accurate information.

Three more layers can make filling out forms in general even worse. The first is if it's actually an actual piece of paper, and the paper is lost. Have you ever completed forms that had more than five questions and then had to rewrite it in order to find it? What was the experience like? How was it the time it took to find it lost, even though you believed there was a team doing work on it for the last month?

The next is the time when the person who requested the information is not able to answer any of the questions about the form. Resumes in general in addition to the mandatory questions on disability and equality can be put in this category if someone has the resume. Another instance is when you're requested to provide addresses, several phone numbers, or Social Security numbers. There are two reasons why this information needs to be taken. One is when someone copies an existing form and believes that the form should have the required information, or they simply decided not to take it off. In the second, they're gathering the data to be capable of selling it, something that happens far frequently.

The final area is where the issue or the data to be completed isn't clear. The location where this happens the most frequently in the business world is when departments are in conflict. An excellent instance is when teams work on a plan or a business case. The words used be confusing. If there's an issue with the language that is a problem, it could be more difficult. Definitions and examples can be helpful but the training process is more effective.

We're specifically interested in calling the time card, paychecks in paper, along with expense report. There's no reason they can't be automated for companies with over 25 workers. The amount of time and money spent in the process of implementing these is astounding. If your time-management system isn't being automatically calculated and direct to payroll, you should implement it right away. A majority of businesses discover it is less expensive with regard to banking costs using direct deposit for payroll.

The same is true for the expense reports. The levels of approval are simply wasted time. It's not worth the effort because any significant thing you could find. It is possible to set parameters to trigger unusual activities that can be reviewed manually.

If you suspect you've got problems with forms or paperwork, as most companies do, deal with it. Don't be scared of it. A lot of people are scared because forms and paperwork are typically the result of an entire system constructed on the top of a system that was constructed on the top other systems. It is essential be able to cut the chain.

An idea for forms could be to set up rules that will be in place. The most important one is that there shouldn't be any paperwork that isn't legally required. All other documents should be digitally-only.

01. It shouldn't be included in a form when we're unable to record data in a uniform manner, as with the ISO document.

02. Make sure that every form is updated to include the minimum amount of effort required by the person filling out it. If you are able it is possible to ensure that filling out the

form is an exchange between the person being asked to fill it in as well as the person who needs the details. The person who is asking to fill out the form, and not the one giving the information.

03. Every form should go through a process of review and approval every year. It must be approved by someone who has not previously approved the form to ensure that fresh eyes can take the form.

04. If it's worth having the form, it's worthwhile to create an explanation video. Develop the training form before publishing the form. The training video should start by describing the value the person filling out the form will gain.

05. Limit the amount of documents your company may be able to. This way when you reach the maximum amount and require a new form then something else needs to be discarded.

Systems Work for You

"That's the way systems function. Sometimes you're the bear, and other times the bear takes you."

Sue Grafton

There are two fundamentals in psychology that we should explore that are explained effectively by the psychologist Dr. Robert Cialdini in "Influence the Psychology behind Persuasion." They're the principle of contrast and resistance to changes. Knowing these principles could be an important factor in understanding the reasons why you aren't able to move forward in various business areas.

The principle of contrast shows that we can notice a greater distinction than when we compare two things with substantial differences at the beginning. Our brains are able to make the contrast more attractive and will have an unintentionally influenced our choices that are influenced by our natural bias.

A good example of the concept of contrast can be found when people look for a new job. If our bias is that the current situation isn't good We will overstate negative aspects in order to make the job seem more appealing. It is possible to use the current salary of $25,000, and an offer of $35,000. The $10,000 salary is a substantial change, but we'd include words like modest to explain the salary of $25,000, or perhaps generous to describe $35,000. You could also consider using the percentage of difference to make the sound of $35,000 better , since it's the type of bias we'd like to have.

Resistance to changes, mostly based on anxiety, is about ensuring an unchanging state. A stable setting, even if it's not ideal to many, is the preferred choice to the uncertainty that comes with changes. We have developed a variety of strategies to avoid changes.

To demonstrate the opposition to change, something is something we often face when we look at the possibility of starting a new job. No matter how exciting the new possibility is there is always a reason to negate every positive aspect. If a possible relocation is necessary, any resistance can be cited as the reason for not wanting to relocate.

Think about a business that's just getting started. They're constantly in changing, which means that the resistance is comparatively low. In general, they don't have many differences between an upcoming service compared to the way they operate. They pick parameters that will enable them to start bringing the most revenue feasible.

If we look two years in the future with that company there is a sense of stability beginning to appear. Revenue is becoming more predictable and the growth is occurring. But, in the present, tension and aversion to change are gaining the impact of evolving systems. We're so focused on trying to expand that distractions like the infrastructure we have and systems supporting it becomes, in the end less significant in the event that they don't impact the next month's revenue.

We can say, "We don't have time to completely overhaul the payroll system, and then tie it back to an accounting software. What can we do to fix this?" We start to develop systems that are built over systems that are based on patches, and, eventually, systems. We're not creating a useful instrument for operations. We're building an institution for ourselves. We are enforcing our citizens to be part of an organization rather than one which is working for us.

When the time comes to think about replacing a system, the effect of contrast will be evident. The differences will be widened to ensure the system is protected. The opposition to change will be based on cost planning, education, and expense to prevent the change from occurring. These are legitimate considerations to be considered in the right context when making plans.

There are significant distinctions between inside and outside requirements for systems. It might not be feasible to alter an industry standard and regulations could require you to perform certain things. But those rules are usually small portions of the systems used by most companies. If the system you are working on needs to be changed, and you're trying to establish yourself as a leader take charge of the to make changes. The outsiders may be resistant however, when things are changed then you're the norm and, in a way you write the guidelines.

Companies that employ LEAN or Six Sigma aren't strangers to system optimization. These methods are designed to eliminate the waste (time materials, time and so on.) and improve reproducibility. They insist that the system must be in charge and the individuals should handle the exceptions and instead of the other way around. A sign that you require changes to your system occurs when you make a system more efficient to treat the symptoms , instead of addressing the issue.

Human nature is to be averse to this actions of systems against systems. We've already bought something and are looking to squeeze the most we can from it. The issue is all the little extras that go into putting the old technology. In the near future, we'll are stepping on one dollar underneath a brick to collect the dime that is dripping in the puddle, because that's the way we've all the time.

It's not saying that this only happens in the business world. Consider the places in your home where you might have something similar to this, and what feeling you might experience when thinking about changing the situation. Do you become like a lot of thought-provoking thoughts are generated from changing one item? Do you recognize that breaking it down into small steps can help and help you feel happier about doing something that has immediate value in your own life?

The plan and strategies should include modifications to your system. It is essential to have separate funds set aside for these adjustments so that the issue of budgeting and funding cease to be a matter of debate. Training and planning are crucial for any system to be properly implemented. Included in your plans is a means for the system to be a priority and be the focus of thought-provoking internal innovationas we will discuss in the next part in this article.

The founders may have difficulty with changes to the system. They have proven their worth with the system that is in place. It's hard to doubt that their success. But, founders possess the ability to give back to their organizations as well as themselves. If you ask them, "What would you do in the future if you were forced to begin again?" typically, they shift their mind. They wouldn't start from scratch since they know. The majority of them would make a range of system modifications. They may realize that their systems are restricting what they are able to do and realize it's time to invest in improvements by making this tiny frame of mind change.

There is a famous quotation from an unidentified author that declares,

"Growth is difficult. It is painful to change. Nothing is so painful as being stuck in a place you're not a part of."

This is something that companies ought to be aware of. As does a quote by C.S. Lewis which states,

"You cannot go back and alter the beginning, however you can begin from at the point you are in and alter your ending."

The larger the company the more difficult this task is to achieve. The impact of changing the system can affect many people in a short time. The change in the process could result in mishaps or chaos if the people involved are not aware or aren't trained. This is why the importance of focus and prioritization becomes important.

It's important to mention that systems don't only concern manufacturing, accounting and information technology. There are systems that deal with human resources as well as management, too. The management systems that surround people have to be free of restrictions to ensure that organizations is able to grow quickly. Individuals make up the company and require the greatest care that is as individual as is possible.

The number of systems are in operation should be carefully assessed. This isn't based on the perspective of LEAN or cost-per-unit perspective, but rather from a standard methodology. 80percent of what your business does must be included in 20 percent of the systems. These systems should be recognized to all employees of the company. In the remaining 80 percent, systems must be centralized by a department, and possibly the location for rules. If this is done with discipline, you'll usually encounter the customer Resource Management (CRM) system and a system for human resources/payroll and an accounting system an enterprise management system, and manufacturing systems.

Take a moment to pause before moving on and consider individuals and systems. Do you ever think it's impossible to separate certain people in the grand scheme of things?

Systems are a method to give people an excuse or cover up. It is used by people to create a bottleneck since there aren't enough people who understand it or are in a position to challenge the claims of an individual. When this happens, it's an area that was once of significance, but the business has evolved. It is still common for people to doing things because that's how we've previously done things. Think about this real-world example of that, and there are many like-minded examples available on Reddit.

"A person is about to retire and must instruct a coworker on a report runs twice a week. The report takes about an average of eight hours to complete in the event that there are no issues. The report's information is intended to evaluate some of the organization's Key Performance Metric (KPM). The person who is retiring will have enough

time to teach the new person how to do it ,
and be observant while the new person is
doing it. The transition goes smoothly, and
the course is successful.

The person who is running this report is doing
it exactly according to the manner they were
taught the previous time, while everyone
pays attentively. The manager is satisfied that
the new employee did it correctly and
communicates with their boss about the
report. The report generator records exactly
how they were instructed and integrates that
to the Standard Operating Procedure (SOP).

When they next execute this report, they
spend an hour to program functions using a
program they found online for free. The
report generator needs another hour to
create all the other information and to ensure
that the report is of good quality. Next time,
they will only require an hour to complete the
smaller pieces of data and quality-checking of
the report since their automation was
completed. They were able to reduce an eight
hours task to one hour. In the following two

years, the team will refined the process to take just about one minute to run to capture all the data and take around 14 minutes to review the documents.

If the report generator goes on vacation, other people have in order to execute the report. People who are following the SOP claim that it's well-written, however they don't understand what the generator of reports can manage to finish it in only eight hours, since they can't finish it in 10 hours.

It's true. Report generators didn't inform anyone about reducing the time to take 1/32 of the amount of amount of time required. The report generator makes use of the remainder of the time to accomplish whatever they like. 10 years on, and the business is still convinced that this employee is very dedicated and is proficient in the system.

It's not the only disturbing aspect of this report. The report is only reviewed two times a year. The first time is during an annual Board of Directors meeting and at the time of an annual ISO audit of SOP. At this Board of Directors meeting, it's simply a number on a slideshow which isn't even talked about. When it comes to the ISO Audit, it's simply an affirmation that indicates that the report is present. The audit doesn't look to see if the data."

We're not going to defend the report's time or the wasting of money. They could have increased the amount of savings needed to run the report through an upgrade to the system. We're not going to be defending the company for allowing time of a worker to be wasted on something that seems to be insignificant. The report generator was optimized for something that really shouldn't exist.

There are numerous questions about this scenario. Why did the business initially required the report? What happened that made them not have to use it anymore? What made the employee diligent enough to improve it however, they were so unhappy that they wouldn't discuss the idea? The reason is that systems, as with everything else isn't set to forget. They must evolve to keep pace with the changing needs of a business.

If it's not broken, then why should you fix it? Everyone wants to make the most of the investment we make in any thing. This can make us forget the fact that we're investing money in something that is no longer of any value. It is a challenge to figure out whether something is still relevant and may appear to be offensive if it's an employee's responsibility. You may also find the need to spend money to upgrade a system which might not fit into the budget. All of these are strategies to keep the same system and achieve similar results profit and variations that remain in place.

Individuals Do Things, not Companies

"Everybody's geniuses. However, when one judges a fish on their ability to ascend trees and live all its life believing it's dumb."

-Albert Einstein

Spend a few minutes and make an image in your mind of a garden with flowers. The aroma of many different flowers is evident before you even get into the atrium. There are many visuals, textures and climates as go on your journey. You travel between space and space, noting the intensity of light, humidity of the air, and the temperature changes. Then you notice huge areas that aren't in construction. You inquire with an attendant as to why these regions aren't growing plants. They answer, "We planted the leftover seeds from other areas, however they weren't able to grow."

This is an extreme scenario that isn't feasible in the botanical gardens. The gardens feature a planned arrangement of the settings aimed at maximising the growth of the vegetation in areas to ensure they can flourish to the visitors' advantage. The gardener will shift the seeds into the right place or alter the conditions for seeds. They'd ensure that every kernel was in the correct spot and not just take the leftovers to fill an empty space.

Why why don't we make an effort to apply the same level of attention and care to ensure that we've created the best conditions that allow people to thrive?

There is a common belief in the world of business that the sole individual responsible for the success of any person themselves. In reality, it's on the individual to find an avenue and perform the majority of the work. However, if you want them to be their best, then you have to assist them in getting there. It has to be about them and the role they wish to fill in your business. If you aren't investing in others, then why should they want to invest in you?

The Golden Rule has applications everywhere. Be kind to others the way you would like to be treated. It's so basic to humanity. If you want to see others put in their time, energy as well as their thoughts and actions to accomplish your goals, then you must take the same approach for them. Be sure to take care of your team and they will grow exactly like the plants in your garden. Sir Richard Branson, of the Virgin business collection, famously declared in 2014:

"Train your staff well so that they are able to quit. Make sure you treat them with respect so that they won't be tempted to leave."

By taking good care of your employees, you're showing how you expect to treat them as customers and one another. In reality, we believe that when you treat your employees as customers, you'll have a great success. Everyone will feel respected and will be able to accommodate. There will be a feeling of caring for each others by listening actively and active taking action.

Does this mean that you shouldn't end relationships with certain individuals? There are occasions when you must take this action.

Sometimes, it's advantageous to the individual in the event that they've outgrown their role and cannot be transferred to a new position in the business. One thing to think about is did you treat as a client and told them that you've all you could done and let off your ego and give them worth?

This applies to customers who buy from you. It isn't always the case that the customer is right There are instances where you need to remove customers. Any transaction should focus on the value to everyone that are involved. Some customers offer little to zero value for what they receive for their money. This realization can be one of the primary aspects of creating a plan to help your business grow. It's a difficult step to make, particularly in the case of a publicly traded business, but it is a necessary one.

Do we have any particular considerations for outstanding performers who also play in teams? Are there any considerations for those who have ideas, motivation, for other people and a yes, and attitude? The answer is definitely yes and be demonstrated. An event that is public with awards and certificates is a great starting place, but a cash reward should also be added.

If you don't take action, either of the following is likely to occur. The first is that those who do be disenfranchised and possibly angry. Then, they'll cease to put in the work they normally put into their work, and eventually they'll leave.

When you're dealing with a truly amazing team member, not like our report generator from our previous section, who has informed you that they're going to leave it is important to consider the best way to keep the person. The first step is to acknowledge that you were not paying any attention to their disappointment when their reason for leaving is related to work. Sometimes, people quit for personal reasons, such as a sick family member or moving to be with an ex-spouse.

If they do tell you something they aren't satisfied with, make changes to the situation.

Actions must be quick and swift. This doesn't mean three or six months. It's a matter of three days, there's plans in place, and after two weeks the changes are being implemented and completed as fast as it is feasible. We suggest brainstorming with employees who might be able to leave in order to advance the process. They may realize that staying in the company isn't a good decision for any business in the event that ideating doesn't produce results. The employees, as well as others in the organization may conclude that you have at the very least, attempted.

You're actively demonstrating that you place people first. If that's what your company claims but then demonstrate it through the real world. The first step for any business who wants to be a leader in technological advancement is to investigate every avenue to keep an individual who is talented. People who are talented tend to be flexible and easily coachable. This means that what they've accomplished in the past is evidence that they can do but it's not a limiter to what they are able to accomplish going forward.

There are instances where it's not business sense to change the way things are done to maintain a person's status? Yes and for an ingenuous business that follows the guidelines we have laid out, these are the exceptions. Finding talented, flexible employees who perform is very difficult. It's costly in terms of the time they are lost in leaving to recruit, train, or leave. In the equation of cost is the impact on morale of the entire team as well as the message received when people see talent leaving the company and there is no apparent action taken by the organization to keep them.

In a moment, let's look at pay and benefits in general. If your business reports records and all employees receive basically the same amount or percentage of increase, you're not treating employees well. If you wish for employees to be incredibly successful in helping you earn money, then pay more than they're worth. This is a sign that you are at the top of your game in terms of pay. If you're leading the market by introducing new technologies that is, you'll be charging customers a premium because they are getting greater value. The premium should be reflected in your staff's compensation as well.

Henry Ford, in January 1914, earned an average daily wage of $5 for employees. Before, they earned $2.34 per day, on average. He also reduced the typical shift from 8 hours to 8 hour days per week. It was referred to as profit-sharing however, he should have named the program enthusiasm sharing. Ford changed to having among the lowest employee turnover rates , to one of the top. In the first year of operation, productivity was up by 40%, while profitability rose by 20 percent. The repercussions of this in the years immediately following saw continual growth in profits.

It is suggested that every year to review the latest capabilities or added value has added. If someone has improved in their job by 10 percent, they should be paid an additional

10% of pay. If someone has taken on new duties as a supplement to what they were already doing then they should be awarded additional money. A 3% increase every year will cover the cost of inflation. If an individual has truly not made any progress or created value for the organization, your initial inquiry is whether or not the organization has invested in the individual to grow.

Be aware that management practices that are popular of having employees perform self-evaluations, followed by the manager conducting an assessment based on the employee's self-evaluation comes at an end. This kind of performance evaluation method is where inefficiency and complacency as well as actions that do not focus on prioritization are exhibited.

If a manager isn't able to give you what the three most important things an employee has accomplished in the previous quarter that directly relates to the priorities of the business without having the worker ask. In this scenario you're in a position to determine the person who is over-working and doesn't have the ability to concentrate on their team sufficiently or simply doesn't focus on their team in any way.

Does an employee have the right to make points that the manager did not notice? Yes they should be able to do so and be capable of registering it in the system that holds the data collected. The key here is that active management doesn't require employees to conduct self-evaluations. The leaders who are actively engaged have a clear picture of what's going on because they're empowering actions to ensure that employees are working towards advancing the company's goals. Insuring that they are active managers will be the second element that requires immediate improvement in communication to shift the culture towards an innovative, strategic approach to the market.

We don't advocate that you do anything in the form of micro-management. It is a major inefficiency for everyone's time. The Dr. Ken Blanchard and Dr. Spencer Johnson do a amazing task in their publication "The One Minute Manager" laying out the steps to take in this manner in a timely manner, and so we will not overdo the subject.

Here's a final idea for the new employees that are joining your company. Let's suppose that a new hire appears to be a great match. The annual salary you would like to pay them is $120,000. Note that we didn't discuss an earlier salary since we believe that you're paying a premium to attract top talent.

To ensure that you are receiving a candidate who is the best for this job Put them on the 30/90/180/180 plan. It is typical for them to earn an average of $10,000 per month. However, under the 30/90/180 program it is paid tilted towards the end of the first year.

In the initial 30 days the salary is $5,500. The following 2 months, the salary will be $7,000 every month. The following 3 months of employment, the income amounts to $8500 per month. After six months of work the employees have earned 45,000 dollars, which is 70% of the possible earnings had they received each month $10,000. If they are performing in line with expectations, you will give them $15,000 as pay, not as a bonus, and the monthly wage increases up to $10,000 per month. They will still receive the $120,000 premium for the first year and constant increases during the initial six months.

There is a chance that you are beginning them at an amount that is lower than what they currently earn but in the end, you'll exceed the amount. People will realize that the system is designed to really reward them when they achieve. People who are truly successful won't view the lower initial rate as being significant since they'll be confident about performing what they claim they are capable of. They are the ones who have the potential for breakthrough Thinking(tm). They may appear good on paper or appear well-presented in interviews but lack the capacity to perform usually be dismissed.

Chapter 6: Belief Over Facts

A wise man once stated, "Bees don't waste their time explaining to flying insects that honey is superior to shit."

-Anonymous

"I believe" is one of the most frightening words you can use in any context and especially in the business world. If you take them as a whole there's no amount of evidence you can prove an assumption. It is not intuition that makes a belief. If we believe in it, it's an act of faith that has to be overtaken with the help of the individual usually by a very profound self-discovery. We must be very clear, in no way, form or form is this discussion primarily about religion or politics.

There are times, particularly within the English language, when words like "I believe" or "I believe" are utilized as a substitute for "I consider." These situations are much better than the latter because they can be a sign of intuition. If somebody has spoken one of these phrases in public particularly in a large number or to someone in higher management, can be a source of concern. They've put a statement in the public domain. If they were to retract it, this would be to prove they were incorrect. Humans have not just a desire to be right, but also a deep need to not make mistakes before other people. It may seem like the same thing but it's quite different.

If we're concerned with being correct, it's more to do with our self-image as well as the ego. If we make a mistake before other people, it's an internal reflection of what we believe others consider us to be. It's not really a perception everyone else has. This perception is also an impact on our ability to make poor choices and more about the subject below.

Three quotes provide different perspectives about facts and beliefs which can help us gain new views. It is essential that we feel empathy , or even compassion when we interact with other people to keep our perspective intact when fighting beliefs against facts.

"Facts don't disappear since they are not heeded." Aldous Huxley

"Beliefs don't change facts. The facts, if you're being reasonable ought to change your opinions." --Ricky Gervais

"A lie isn't a truth, wrong isn't the right thing, and evil doesn't transform into good simply because it's accepted by the majority of people." Booker T. Washington

In business, often when the decision is made the majority of people believe that it has to be the best decision because there's no turning back after the decision is taken. Two aspects are able to be changed to change that mentality.

The first step is to recognize any decision you make, you have to keep making that decision every day. The issue is that many people think that every new day is exactly like the day you first took a decision. But it's not. If you find that something isn't working and you've gained a new knowledge, it's acceptable to alter your original decision.

The other issue is acknowledging that in public we have made a statement. The issue of being right can hurt us because it contradicts the way we're wired to function. We'll go to extraordinary lengths to defend a wrong decision in order to avoid embarrassment and then risk losing many dollars to the consequences. Pride is the most destructive thing humanity can do.

We allow ourselves to be a bit naughty by using "I believe" or "I believe" because these phrases are very personal and intangible. The language allows one to be right regardless of other factors. It is difficult to prove how an event affected someone's feelings or what they choose to believe. It is possible to have an effect that is compounded if someone has made similar claims previously and has had good results with the idea they've said.

The expression "I believe" is a bit more flexibility in the context of a given. The phrase implies that a thought process that is of some kind is taking place that is based upon a data established. If we introduce new data that is based on new data, then new thinking may arise.

If you're a manager in your organization and you are the leader, ask your employees to highlight you when you are using any of these sorts of phrases. They're trying to get them to show awareness that you're indicating without intention that you're unwilling to make decisions based on data. By doing this, you're indicating that you're open to new information and are determined to make the best decisions by using data. It eliminates the perception that many people have that management of a company doesn't take note of what's happening.

To respond to someone who says one of those phrases, say "Thank that you shared this. It seems like I have to think about different perspectives. Could you give me the three main reasons why you feel or believe this way regarding XYZ? I'd like to conduct some study and determine if my information is incorrect and discuss it with you in the near future."

It could require a few instances of trying this method before the culture is able to accept the idea that you can do this. When conducting your research, be honest in your efforts to discover an alternative view of the information. When you do this you can are able to appreciate the person by pause and examining their perspective. You are willing to acknowledge you may be incorrect. You show that you are eager to know more. Then you make it your top priority by making it a point to get in touch as fast as you can.

It is possible that you did not notice certain data points and you could be the one who changes. This is known as growth.

You must keep an eye on the war that is going within their heads subconsciously. This battle could result in their not saying anything or thinking in a disconnected manner. If this happens, say, "OK, I would like it if after this conversation we could discuss one-on one and you can give me the information you have so that I can comprehend."

If there are other people present and you're displaying an the willingness to hear other's opinions. Also, you are demonstrating that you don't intend to place the person in a position to be a nuisance in order that the discussion can proceed.

Certain people won't be interested to try this strategy. If it's the CEO making these kinds of statements this is extremely harmful for the business. It may be necessary to determine how to bring the rest of the upper management to agree or to let the issue go.

We hope that the people and companies reading any chapter in our Breakthrough Thinking(tm) series are open-minded and are looking to think outside the box, not repeat. It's easier to avoid the flies in the quote in this chapter if your business does not.

Emotional Intelligence and Awareness

"I've discovered that people forget what you've said. people will forget the actions you took however, they will never forget the impression you made on them."

-Dr. Maya Angelou

"The Other Kind of Smart," by Harvey Deutschendorf is an excellent book that breaks down Emotional Intelligence and gaining a greater understanding of yourself as well as other people. The excellent tips and advice will positively impact your performance across all aspects of your life and work, not just business. We'll limit our discussion to the ways that increasing your emotional Intelligence collectively allows to develop a breakthrough Thinking(tm) approach, however, we would also suggest reading Harvey's book.

The core of Emotional Intelligence in our purposes is to be aware. It is about you first, and then the other people around you. If we recognize our own biases it is easier to control the things we do and how we are focusing.

The ability to control our perception allows to be more open and a greater sense of compassion towards others. Being aware and investing in the other is where the magic of groups takes place. It could be best summated into the notion one of the most important things you need to recognize about any other person is that they're not like the same as. All else is secondary.

Seven influences affect our lives every day. We're clearly not taking one of them to be either positive or negative. It's not like the influencers always stare us in the face, neither. Sometimes, they're subtil, and we might not always recognize them. By pointing them out we can bring awareness to the ways they affect your life and the lives of others, which impacts our Emotional Intelligence.

The order was chosen because of the way they link to survival as well as our inherent need to blend into other members of the species.

01. SEX- The core of sex is the concept of survival. It is the ability of a person to endure to the next generation and, consequently immortality. This influence is the most basic level. There are numerous stories of how people aren't under control of their impact and the way it affects their lives.

02. Families and friends- The saying goesthat "you can choose your friends, but is not the same for your families." Both groups can unite the second component of humanity , which is social acceptance. The people we surround ourselves with are important and directly affect how we interact with other people. They influence how we view the reality of society, what is normal and the fundamentals of trusting people.

03. PEERS - The people we interact with are like friends and family. There is an additional aspect of our interactions with peers in comparison to family and friends which can be less formal however, it can impact our interactions. The social aspect is prevalent in our interactions with others and is certainly a

factor in our mental well-being. Acceptance could lead to more bonding, which can result in a greater work-life balance. While rejection is detrimental to teamwork in any way.

04. REGULATION- Regardless of the beliefs of a person's religion, religious beliefs do influence our society. In some regions where it is reflected in laws of the government and on holidays. It is a part of our personal tradition, and is something that must be respected to establish meaningful relationships.

05. Media- What we see, listen to and read, listen to, and listen to influences our thinking. It can open the door to openness , however it can also increase our behavior. The media, excluding religion, can also influence how we are controlled.

06. Money is linked with survival and social groups and even ego. In our modern world we can use money to meet the basic needs as described in Maslow's three-pointed triangle. How much money we own will also determine

the amount of social interaction that will occur. It can directly affect how we interact with frequently. The ego component is evident as a result of the fact that according to many opinions what amount you've got is the determining factor in your life's success. It creates a desire succeed over other people. Also, it can short-circuit the rational mind, which may take some time to make a rational decision.

07. Physical Environment - the physical surrounding environment can motivate or drain us. It is better to perform with clean air, natural light and sound structures, as well as space and hygienic working facilities. Our surroundings could be a signpost of what we would like to achieve. In this case when you ask individuals to act professionally and then put them in a renovated school with chalkboards, you're not giving the right message. Additionally, the space aren't supposed to be perfect and clean in appearance, unless it's an office for medical professionals. It's a strictness of perfection,

which does not work for breakthrough Thinking(tm).

It is important to note that the impact of these seven factors are more emotional nature. Most people make their decisions based on feelings. This isn't likely to change, but becoming conscious of that fact can help you understand how to spot when you're being in the process of being influenced. We invite you to take some time and reflect on what each one of these influences makes you feel when you consider the influences a more thorough consideration.

As you begin to contemplate the yin and yang characteristics that each influence has, you start an acknowledgement process of your own. Awareness is essential to bring authenticity when working with people in order to bring change. Change is a difficult process which requires emotional commitment from everyone in the team to last.

The emotional support by your group will begin to build new memories. The team will start to discuss how they feel about the situation that transpires around them in order to make a change less terrifying, but instead an experience of joy. The excitement of change will naturally result in a better experience for your customers when you begin to truly create new markets.

The experience of customers is becoming more crucial than ever before in relation to customer service. If you plan to treat every person as a potential customer, the ability to recognize Emotional Intelligence must be a top priority when it comes to preparing your company to transform.

What is the key to success?

"What is easy doesn't last, and what is durable isn't easily."

-Anonymous

In the numerous live sessions we've held during our time at the Ideation Emporium of Creativity(tm) Intelligence is frequently cited as the most valuable attribute an individual has that determines the success of their business. But it's not. Intelligence isn't even among the top 5. Being able to stand up for yourself, refusing to be a victim to someone else, communicating at a higher level, being humble, and being able to discern when to act are the primary five qualities that are frequently mentioned to determine the odds of being successful.

01. GRIT - The discipline of staying to a particular idea is the most important quality of a successful. You never give up on the things you believe in and overcome the challenges that hit you. John C. Maxwell outlines the Law of Consistency in his book "The 15 Invaluable Laws of Growth," which provides the most in-depth explanation on this crucial factor of the path to success.

You must overcome your own neuroplasticity, as well as the distractions that afflict you from the other. Once you begin to concentrate and concentrate, you'll have to remove other things intentionally. Stopping doing the things you previously did cause a shift in others as well. They might not even know about it but likely will attempt to get you back to the way they are familiar with. In this moment you'll need to be disciplined to keep doing what you've started.

Staying with what you've started is an easy idea , but it's not an easy one. The most important reason why people don't develop their ideas effectively is because they're doing too much and are not completely committed to what they are doing. They're constantly thinking about the possibilities instead of rushing to create the 'what should be. At The Idea Center for Creativity(tm) We recommend that after you've committed resources to make any move do it only if you can prove that your alternative that is 40% or more superior.

Our goal is to boost your capabilities to innovate, such scenarios rarely occur. If they do occur they are usually large in scope such as an epidemic, but aren't likely to be a good idea to plan for. But the difference is that you'll be able to change, with discipline to improve your clients and their lives, while others are sucked into anxiety. They'll talk about the past returning to the way they used to be. It is evident that this isn't feasible, let alone is it even possible.

Do not confuse having grit with doing more. The aim is to be more efficient to ensure that you remain motivated when work becomes challenging. It's possible to have to be a bit more gruelling but this doesn't mean that it has to be negative if you're also doing your best.

When the most elite military consider whom they'd like to put for their team They select those they trust but not the most effective. Of obviously, if you find a top performer who is also extremely reliable to a group and is as reliable as it can get. The most important thing to remember is that a moderate or less efficient player who can be counted on will be a better fit in the most successful teams all over the world.

A person who is focused, who can remain focused and conquer the challenges consistently is the one who wins. They don't need to be the strongest, fastest or the smartest. They are able to rely on their team to deliver every time. It builds confidence that members aren't immediately going to be stricken by the desire to go about it in their own way, instead of following the established strategy. The team works by focusing on the strategy.

Then, grit, once you finish something, brings satisfaction. You'll feel empowered and encouraged by seeing things from a different perspective. You'll find the self-confidence you've always wanted that can bring you happiness.

02. Saying no is a part of discipline. It is the obligation for leaders to say no. Ideas that are

good are just like a bus in the city. There's always another around the corner. If you decide to commit to a plan, market and put resources into the idea, you must say no to any other suggestions. Saying no doesn't mean never. It's an indication that your team is focussed and focused with what they have dedicated to in order that you can accomplish it speedily to add added value to your market. New ideas and projects get added to the pipeline once things are implemented.

The inventive ideation and ingenuity process are explored in other books of this series.

People aren't averse to saying"no. They feel that they're denying others. It's not true, you're actually helping people by not allowing them to waste time. We often think that"no" is negative. Take a fresh look at it simply being an option.

03. Creative Communication - As we have discussed in the previous chapter, communication is essential to bringing groups together and is the top problem that many

businesses have to deal with. It may be helpful to review the five C's: calm compassionate, kind as well as concise and consistent and again, as well as the three major talking points that were introduced in chapters 3 and 2 respectively. If we are able to add the spark of imagination, particularly in communication, we are able to excel. Most often, this is due to the desire to be creative, however it could also be due to the individual messages that touch on a personal level.

04. SHOW HUMILITY - The power of Humility starts by being HUMan. Humans aren't perfect. is the possibility of making mistakes. Recognize when you've been mistaken. Be honest and apologize, and look for ways to amend your mistakes. This sets the tone for others to know when they do make a mistake, they need to acknowledge it in order for the team to take action to correct the issue together. You're demonstrating true determination and strength by doing this.

This act encourages the exchange of information and builds trust. The people don't have to be shy from the world anymore. They can beat the fear of being isolated by recognizing their flaws and how these imperfections make you more attractive and not being a snob.

The other aspect of humility is to recognize the work of the team. Find people who are who are doing the right thing and then acknowledge the effort. Recognition of positive behavior is more difficult than it appears. Let people know that you appreciate the amazing things they are doing and, most important, truly appreciate them. In this instance the word "us" means the person and the result from their activities.

We tend to be able to go along even when everything is going smoothly and just look at the bumps. But, to achieve breakthrough Thinking(tm) it is essential to concentrate on the positives to ensure that, when something negative arises it can be dealt with by the team it quickly without anger that can lead to burnout.

Everybody can identify those who take credit, but do not contribute significantly to the project. Give credit to teams and celebrate them to those who deserve credit. Also, you should be celebrating yourself. It's not a sign of humbleness to not acknowledge your work. It could lead to a diminishing of your worth, which can lead to damage to your personal image. You must join in the celebration and bow with the rest of the crowd make sure you perform that.

05. LEARNING TO FOLLOW- Leading isn't about being always the right way. You won't be 100% right all the all the time. Being a leader means having a team of experts who can guide and grow. Leaders know when to allow others to follow in the direction of the leader to help a team achieve what it requires. The leaders who are able to effectively follow the leader's instructions will show to the group that they are part of the team and not only the team.

Flexibility serves two additional vital tasks other than the ability to empower others.

First, it shows the trust. The trust that an organization has helps to keep the organization on track when distractions pop up. The fastest method to earn trust is by granting trust. Like respect, genuine trust can be earned through time, but you must get the ball moving.

Another benefit is helping to develop leadership qualities among other people. They can learn from your guidance along with other leader. This doesn't mean that they won't likely make mistakes since they are bound to. What do you think is more important: that they fail, or F.A.I.L.? If they fail the are a disaster and they melt. They lose confidence and might not understand the reason or how they've been unsuccessful. In the event that they F.A.I.L. (First Try In Learning) it is possible to get others' perspectives to identify the things they excelled at. They can also be open to the idea that there could be an opportunity to do things differently.

Reminisce about a moment at home, when you needed to step back and observe someone else do something you had taught them. Did you feel confident and confident that they had mastered the job done? Then, turn that reflection around. Was there a time you were instructed to perform something, and then assigned the task of it? When you succeeded in completing it, perhaps you heard anyone who was either the one who was instructing or guiding you to tell you how they were proud of your accomplishments.

Pareto Principle

"Feeling unhappy or sad, or upset after making a decision isn't a sign that you've made the wrong decision."

-Anonymous

Vilfredo Pareto probably didn't think that his observation on the distribution of wealth across a population to become so widespread and widely used, but we're grateful to have it. This pattern of distribution is called"the sparse principle," also known as the law of the essential few, and more frequently called"the 80-20 rule. It can be applied to economics health, safety, sports and quality control, triage and SKU (Stock Keeping Units) sales. A good example is it is that 20% of SKUs make up 20% percent of revenue. However that 80% of SKUs comprise 20 percent of the total revenue.

The principle also applies to the way people spend their time. Let's suppose your company runs using a five-day schedule that includes eight hours of working day. It means that on an most days, one of the week , you will be able to complete an average of 80% of your work.

The first reaction you might have is to consider how ridiculous that assertion is. It's possible that you're correct. Your team or organization could be among the 20% of companies which don't suffer from this problem. This is the test. Have you ever thought several times during a month, why it takes an inordinate amount of time to accomplish tasks in a firm which is proclaiming that they are innovating? If yes, you may be part of the group is in this category.

If you spoke with people regarding their daily routine, you'd get a myriad of different stories coming out. To get out the noise, you should inquire about which three things are their most important priorities and how are they linked to the company. If they have trouble answering the question, they're too disorganized and are pursuing the wrong targets. But, that's the past and you can make it better.

What happens if you modify the equation to 80percent productive on immediate priority? You get exponential results. If you have all employees working on prioritization and empowered to block out the noise, the result of 80% isn't just 4x. It's actually around 10x and in certain companies it's more than the four-fold increase. According to the research, employees are achieving and feeling fulfilled from the outcomes that result through their effort. This is bringing happiness to the natural environment.

The remaining 20% of the budget should be set aside for employees to expand as they explore and grow. The goal is to ensure that the productivity of 80% is not impacted to the reverse. It is an investment in preventing burnout.

Furthermore, if you're investing the 20% in individuals and there's an emergency, or a genuine emergency, they are able to handle the demands of the emergency quickly. When we build facilities, we aim for 50 percent, 80% 90 percent, and 100 percent timespans if we require capacity in the unexpected. It is important to plan for the people in the same manner.

If you are worried about having people who are sat around doing nothing there is already the problem. Take a step back and think on how much time and energy you think you spend on things that aren't important. Does it give you a feeling of fulfillment or bring you joy? Most likely, it doesn't since sitting around isn't really entertaining, it's just a terrible method to relieve stress.

The employees won't just sit in a circle if you alter the structure of work because they'll be working on issues they are eager to tackle together with you. They will be set up for success by giving them the ability to empower of time, pay, and investment.

In general the 80-20 rule can make a fantastic starting point for tackling the majority of concerns. Even in "Miserable at Work" Why? It's Not Required to Be.(tm)" We explore how this rule can be applied to the ideation process for expanding and decreasing concepts. It is often possible to discover new perspectives about time and money when you start by asking "If the 20% portion of your efforts yield 80percent of the results What could be changed to the remainder of the work to improve the outcome?" The reason for this issue is that 80% of the time, effort is spent without generating any output; this is the reason burnout and fatigue can occur.

Chapter 7: Prioritization

"No business can do it all. Even if it has funds, it won't have enough competent employees. It must set the priorities. The most unwise option is to do doing a piece of everything. This will ensure that nothing gets completed. It's better to choose the wrong priority, rather than having none ever."

Peter Drucker

If you are struggling with prioritization it is important to know that it's a common problem. One of the most important reasons for writing this book, and in this entire series, was the amount of times we get this feedback from our colleagues in live sessions of the Idea Emporium for Creativity(tm).

In the final chapter of this book we expect that the message on concentration will be well received. There are times when people are simply doing too much , and not doing anything to make progress. When you concentrate upon Breakthrough Thinking(tm) you will discover prioritization becomes simpler.

The method we use to determine our priorities is fairly easy. The first step is to determine the market sector you're looking into. It's typically determined by the company's previous experiences or in the case of a new venture, on the management's experience.

You must then determine the specific area you're seeking. It isn't universal. It is not the way to go in the true sense of innovation and you're fooling yourself to try it. Start with one location first, then try it in the next area of the globe. The majority of the time this is the place the place where your headquarters are but it could be also be based on customer locations , or maybe the issue you're solving that is geographically based.

After you have identified your geographic and sector targets then your team can begin looking at potential projects. We evaluate the appeal to the market and what speed the project is completed. Both attractiveness as well as speed of a project are placed on a scale from 1 to 10, and plotted against one another. Attractiveness is shown on the Y-axis while speed can be found located on the X-axis.

The final factor is an estimation of gross profitability or revenue or gross profitability, depending on which is a better gauge of the market/business we're analyzing. This is shown in a bubble which can be added to the scatter plot we made to measure the attractiveness vs. speed.

Anything that falls in the upper right-hand corner is worth a look. The most well-known circles are those that determine the sequence in which projects are carried out. We've provided a couple of different methods to estimate size of bubbles in the downloadable section. Please visit www.IdeationEmporium.com/downloads

for templates related to this section. Your password will be "innovate."

We provide a few different ways to do estimating for the revenue/profitability. These estimation techniques can also be utilized in the process of innovation as described in "How Are Kids Innovating Faster than You?(tm)". It is also possible to use several methods together depending on what you prefer.

Write down your thoughts and make the best you can guess. If you have reports on data they are helpful however they're also incorrect. The majority of data reports have an error-tolerance of 20% and are highly dependent on the language of the person who is who are collecting the data and the data sources that aren't always fully publicized.

There's a certain art in finding out the best way to make this process impartial. It is an issue where most people would like to have tangible data, but forgetting that, in the case of true innovation it is creating markets. The reason you should start with your first screen of ideas should be required to be located in the upper right portion of your scatter graph. Then, only after that, do you examine the value. The value will be derived in a clearer way with your customers. As of now you're only taking the course, and not the precise route for the journey.

Once you've got this, you can begin with the Voice of Customer procedure, that is described in the chapter of "Miserable Work" Work? Why? There's no need to Be.(tm)" Once you've got that information via your Voice of Customer, you can evaluate different projects.

The most frequent issue occurs when you have large-scale projects that require real innovative thinking compared to projects which are an iterative process. There is also an impact that is compounded when you include internal projects. What is the best way to decide the projects to work on?

When you're more comfortable with the process, begin with three different projects. One is a truly innovative idea that is iterative and the third is an internal initiative. They are the three most important areas for the business to focus on. As you can notice in the area of strategy there will be projects that follow these , too, in order to inform teams about what's coming next.

When you feel more confident, we suggest that you keep no over five different projects in any one moment. Ideally, you'd be able to handle three main projects, and the ability to take on up to two additional minor projects. Focus is the key to success, so you will grow faster by using the power of three and five. There will be instances when things go wrong , and the need forces you to take on a new project? Sure, there will be times when it happens however we haven't seen that you're carrying more than two of them in any given moment, excluding an actual catastrophe or pandemic. There is a chance that a personal project or a new project may take over. It is possible to present them as a matter of urgency, however typically, they're simply an opportunity to distract.

Be aware that the rule of less is more.

After a project has started off, it must continue until it's completed. All new projects must undergo an evaluation which include Voice of Customer. Only if the project is more that 20% even though we suggest 40%, in terms of value, over a project currently underway can the new project start. It doesn't mean that you can't continue any ongoing project, but the resources need to be allocated. This may mean adding staff, but and not expanding the number of staff that you already have.

At any time at any given moment, one of the projects must be truly innovative in order to keep the longer-term plan moving ahead. It is important to continue going forward in order to create profitable and sustainable income outside of the realm of commodities. If you don't, you could easily be caught in the trap of focusing only on things that are iterative.

Iterations are essential to take your business more than the initial 15% market However, they are an opportunity to expand geographically. So, you're going to see a number of these kinds of projects.

It is also a good idea to goal to incorporate an internal project at a minimum every two times an assessment of any potential projects is conducted. It is essential to be aware of infrastructure or building inefficient systems over outdated ones.

It could appear to be a plan slow down your company. It could require some additional time. These strategies could dramatically alter your mindset. These changes will require dedication, time positive reinforcement, determination to produce the desired results. When you begin to implement these kinds of things, you'll notice that you're moving quickly. Actually, it is possible that you might be moving so fast that the timeframes suggested in the chapter on strategy could be reduced in half!

Strategy Basics

"Strategy that is not backed by tactics can be the most ineffective path to victory. Strategy without tactics is the sounds before the defeat."

-Sun Tzu

There will be three areas we will be discussing for the strategy. It will comprise a five-year overall strategy, a three-year sector-specific guideline, as well as a 1 year operating plan. The subsequent chapters will address the details that need to be included. Be sure to follow the order in which they are in the list.

Our ideas may be different to the standard ideas of universities and other businesses do. That's the purpose. We're cutting out things that aren't necessary to make it more innovative. We're not saying that other methods aren't working. If you're not trying to invent, those alternative methods could provide the more details needed to gain your edge over competitors in the areas of commodity that make up the markets.

Strategies are usually paper-based exercises. They could be as thick as a book and after being presented, they can be put aside. It is a waste resources and time, without at the very least a certain level of thought involved. It's more than nothing, but when you're going to invest in developing a strategy tool, you must use it. Inactively using it could be like investing all that money on software only to have it on your computer's desktop and say that you've got it.

To be able to innovate efficiently and keep the focus to progress it is essential to have a strategy. Without a plan, you're simply doing things that are closely related for a certain period of time, and then when the next thing arrives, you become distracted. As a leader, need to establish the direction for the whole group and make sure they are focus.

Start by creating the goal of having a clear, short three-sentence vision for your internal organization. This vision for your internal team is different from what you put on a web page for everyone to view, which is often described as mission and vision declarations. The internal vision should include the region and market that you serve and the unique differentiation that allows you to be a participant in that market and the market share you can expect to gain within five years. A similar approach to the three examples could be suitable.

1. We will lead the way in the domestic climate control market across the United States. Our products will offer four times as effective as self-cleaning filters that remove

contaminants and other harmful substances from the air. We'll expand from our current market share to a 30 percent market share in five years.

Example 2. Within continental Europe We will push the market for fabric that produces power ahead. Our clothes can be charged by any wireless device manufactured in 2021 or later by being within two meters and being comfortable, breathable, and stain-proof. We'll achieve 5 percent market share of the apparel sales within five years.

3. Western United States within desalination will be our main market. We will lead the way in the implementation of self-indicating filters and parts which detect salt build-up pH changes, and substantial pollutant levels, which will make it much easier to build additional plants to alleviate the water crisis. We will be included in at minimum 60% of existing plants, and at least at least 80% of new plants within five years.

The key to staying focused in a plan is to discuss it continuously. The rule we adhere to is that for meetings that are scheduled for Monday, we go over the single-sheet approach around halfway through the session. A five-minute interruption is a great way to focus the group. The quick distractions can help bring attention to the primary subjects. Randomly introducing distractions during the meeting will ensure that everyone hears the messages. Sharing the plan is the most effective method to ensure that everyone is in the same company.

We recommend that at each gathering that includes three or more of management's upper levels that the single-sheet is examined and then be a part of the five-year overall strategy. The objective will not make the information unpleasant for people to hear, however to make it in the forefront. It is important to concentrate on what you stated you would do to accomplish it.

Year-end Overall Strategy

"He who does not plan is planning to fail."

Sir Winston Churchill

The primary designers of the five-year global strategy, which is the first time it the plan should be formulated by the CEO and the Board of Directors. There are three reasons why these people should take on this task. In subsequent versions, you could have members of the Board of Directors replaced with the direct reporting of the CEO. However, the exception is the moment the new Board of Directors member joins they must participate directly in the creation of the new version.

Firstof all, everything that is being formulated comes directly from top. The top executives are laying the foundations for the business, including the sector of business, innovation geographical areas, and internal goals. This is the basis for everything that is. If team members are aware that a document was drafted by the

CEO, and not simply signed off with their signatures, it has importance.

It also ensures that those who decide the parameters are aware of the motive behind the parameters. The intention is important when there are questions later in the scope. It facilitates better integration between departments and operational levels.

Thirdly, it eliminates the black bag operation and pet projects. It eliminates the confusion that may occur when middle management formulates the overall strategy that guides the entire company. The single-sheet overview of the five-year strategy in Monday's meetings also aids in keeping side projects under control.

The five-year plan does require an annual review. This is due to three reasons. In the first, it is likely that you're meeting a variety of the priorities that are coming to the attention of customers. In the event that this happens you'll create market shifts. These market shifts could cause your strategy to change in different areas. In the second instance, if a priority has to be changed that takes more than one year to complete is required, it must be part of the strategy that is changing. Be aware that the impact of the proposed new prioritization should be greater than 20% higher than the value of stopping performing something you've already begun. We recommend 40% and don't recommend the end of something already in the process of being completed, but it is better than adding more work to the plate of people. After five years, it is time to look back and evaluate the places you stated you'd be and the current situation. This is an annual check-up beginning in the sixth year.

We suggest that the first statement, which is distinct from the internal vision of the 5-year plan, be something similar to:

"This is our strategy , which all employees will adhere to as a unit. We're giving you our strategy so you will be able to see how you can contribute to our team. If you're not sure the specifics of what you're expected to complete and aren't sure what it means and how it is a directly and pertinent connection to the 5-year plan It is your responsibility to inquire to speak with us immediately. It is the duty of the top executives of our company to make sure you are aware of the significance of your work every day, and how we are able to appreciate your contribution."

Make sure that you adhere to the assertion or any other you make. It should be genuine and accepted by all employees regardless of position within the organization. It should be the primary item, in addition to the title, that appears on the single-sheet summary, too.

Remember that we're not re-creating the business model that is traditional that is designed for businesses who want to be creative. Your five-year overall strategy should not exceed 15 to 20 PowerPoint slides in length. We're not particularly fond using PowerPoint as a tool for communicating messages, but we recognize that there are times when standard functionality provides an opportunity for completely disruptive. As we explain each slide, remember that you will need a Word file or Excel sheet might be required for assistance. The best rule of thumb is that for an PowerPoint that has fifteen slides in total, you should not be more than two pages of information including tables of numbers, to accompany it.

The overall plan is restricted in its scope. It's important to focus on the key elements that add value to the slide that everybody can be able to comprehend. Remember that in the future you'll have to distill the information into a one-page overview that is shared with all employees within the company. This can be a major issue for some regardless of your company's size. There is a need for people to be immersed with the minutiae of what they've always done, or even what academics have said is required.

If you study a variety of methods, you'll discover that the majority of the data doesn't have any the same value. Unfortunately, the majority of the information collected took some effort to create. If you require an objective test to determine whether data should be added, you can ask three questions. Anyone who inquires about the necessity for the data should have an innovative mind. This is true for finance as well. Finance teams tend to be the most difficult when it comes to change. They also tend to employ a variety of subliminal strategies to keep the status regular.

01. Who has access to this data and what clear concrete, tangible decision will it permit them to take?

02. If the people who asked for the information were to go away and be replaced who was taking their place make use of the information in similarly?

03. Do you want to know if this information is being asked to be used at least once every three months? Can the information requested alter the direction of the business within the next three months?

1- Direction. This slide is your opening statements and internal vision. It is the slides with the most amount of text. It's the only slide that has its entire content on a single sheet summary.

Slide 2: Markets. You can define markets by discussing three factors.

01. What physical regions are you planning to operate in in 5 years?

02. What can you do to lead the marketplace for innovation? What is distinctive about your services that can be explained in less than five words that will allow you to take the forefront of innovations (i.e. intellectual knowledge, expertise, technologies that are adapted to your market joint ventures, joint ventures that don't meet demand in the market that

other companies don't be able to meet, regulatory hurdles such as.).

03. What are the potential markets adjacent in relation to your current solution, two levels down? Think about solutions, geography, and applications. Here's an example that uses the previous internal vision. Example 1. Current: Home climate control within the United States. Level One: Home temperature control within North America and Europe. Level Two: Home control of energy use across North America and Europe.

Slide 3: Customers. Discuss your customer's characteristics by examining three elements.

01. What qualities should a great end-user or customer possess? (i.e. size, quality of their brand as well as their regularity of purchase, and which of them are on the cutting-edge adoption standard distribution, the likelihood of forming positive relationships, overall goal, and the potential

to require many options to them, and not only an item or service). Another way of approaching this is if your business were taken over by a buyer What features would you want for them to possess that go beyond cash?

02. What traits of your customers will cause you to stop selling to them or what would cause you not sell to them?

03. It is important to define what a big customer, medium customer and a small one is in relation to their number of places within your proposed or current geographical regions and the potential annual revenue for your company (not your revenue) and the market share they hold (we will define this in the near future) and their credit score (similar similar to Dun & Bradstreet), the place they usually fall within the cycle of innovation adoption, and an illustration.

Slide 4 Go-To-Market This is a relatively simple slide that explains, from a the source, how you'll use to get to the market, the geographical location, who you'll be targeting in the area and what requirements they may have.

Slide 5: Market/Customer Value Chain- This outlines each process you need to follow from the time an item leaves your facility until the end-user receives the solution and utilizes it. This could include component manufacturers, distributors as well as third-party equipment utilized in conjunction to your solution, logistic firms as well as retailers. It will allow you to understand their how they impact your business and what you can gain from them. It can also help with Voice of Customer, as we cover during The breakthrough Thinking(tm) Series.

Slide 6: solution Road Map- This is the entire lifecycle of every solution. Be aware that as an innovation leader you have the goal to become obsolete before your competition can consider it.

For slides 7-11 Think about three different aspects.

01. What areas do they currently require help to be able to provide for the business?

02. If you reach that goal, how many are the new items will they need to address?

03. What are the specific changes they have to prepare for (expansion or reduction, new headcounts, locations and so on.)?

Slide 7- Logistics/Service Delivery

Slide 8- Customer Service

SIDE 9- Manufacturing

SLIDE 10. Information Technology-

SLIDE 11 Human Resources-

Slide 12 Succession Plan- This page displays the key positions in your company who is learning to be able to take over when the incumbent is promoted or quits. The single page can inspire the desire to encourage from within. If you're in a position that's arises that you believe nobody is prepared to fill that isn't due to a need for a new solution, then your management team should be focusing immediately on the issue.

Slide 13- M&A/JV If you're a company that's innovative it is not wise to look at Mergers and Acquisitions (M&A) to simply take over a competitor to gain market share. It is best to consider this for an integration of the supply chain vertically and gain a presence among native-language people as well as to provide an additional avenue for a solution that could be an innovative idea to a new market. In the case of the last we suggest an Joint Venture (JV) or perhaps a brand new company to allow the parent company to keep its focus.

Slide 14: Growth Summary - This displays the anticipated revenues, profits and major successes as you enter every new year. It serves as a guide towards the goals you need to achieve to be achieved while it is the initial slide to talk about the revenue and profit. To date it's been focused on market share. It's either creating a market from scratch or creating a brand new market by using already existing ones. Additionally, you are bringing new value that no one else has.

Profits or revenue calculation will be determined by the amount of value that is lost. If we are focusing on dollar figures and talk about dollars a lot and it leads to the focus to shift towards money instead of value. Value is the source of money and not the reverse.

Slide 15- Five Year Review - This is a simple slide that's completed in the 6th grade. It compares where you'd be in this moment five years ago with where you are today. Record the most important lessons you have learned, compared to where you stated you'd be and where you are now. In the following year, can compare your the improvement year on year.

Chapter 8: Having Intentional Fun While Working

"If you are doing the things you love, you won't have to work in your entire life."

Anonymous

We've all heard this quote or some variation of it. If you are enjoying the work you do and you are happy with the work. task. This is usually seen as an exclusive privilege for only a handful of people and is not achievable for most. We respectfully disagree and are happy to take the chance to demonstrate how you can be happy in your work and why having fun is extremely essential.

Let's look at the idea that it's good to be happy within our working lives, but that doesn't happen for us. If you consider that you feel like it's wrong however, the social rules that you've always followed say that it's not the case. We must and should be having deliberate and deliberate fun at work. This is the the most crucial element of this how-to guide, and the place we begin with The Ideas Emporium Creativity(tm) whenever we have live sessions with our partners.

Did you have a genuine smile today? Consider that. Have you had one moment when you completely smiled and felt that distinct sensation? The first step is to start each day with a smile in the first half hour. Begin your day by doing things that make you laugh, smile and laugh, or even have funis a way to set the tone for the rest of your day. It will take only a few seconds. Have fun with your children by playing a song that they can perform or dance to. Perhaps go online to learn some new jokes or watch the most popular clips. Find your first smile.

When we're having a blast and enjoying ourselves, time seems to speed up. For those who are more outgoing perhaps it's an enormous party, and those who are more introverted, you can be playing an evening game of board games with some of their friends following an intimate meal. If this is the case, those times seem almost amazing. Most of the time, you'll wish to continue participating in the activities as it's like they've only began. It is odd that we save those time slots for weekends , or when we do not need to work the following day.

The issue of separating the time we should be having fun is our next step. In reality, if you're working in an place where you aren't able to be having fun and you aren't enjoying yourself beyond the money you earn it is time to look for other jobs. We spend so much time with our loved ones to let the burden of. Of of course, if you are to be in prison and are in a prison, you have no option, but you can in nearly every other situation.

It is important to know how to be a fun-loving person again, regardless of where you may be. People will naturally gravitate towards you since people who are fun attract other people. The attraction of that magnetism is strong. It can help you begin developing exceptional teams, or even groups of people that all want to cooperate with you on whatever it is you're doing.

Consider the last home improvement project you worked on yourself, and that needed help from a few people. Did you contact your colleagues to get help? Or did you phone a friend or two? You could have done the same for some which is great. In the case of others, they definitely have friends who they call. We make calls because we believe in them and we want the process to be enjoyable when we complete it. One of the reasons why you have confidence in them is that you enjoy your time together.

The most impressive instance for the effectiveness of humor in attracting people comes from children. Children can be seen laughing between 250 and 400 times a day, as opposed to 17 times per day for adults. Have you ever you witnessed children laughing and wondered the funny side of it?

Their pure joy will make you smile naturally. You might have seen something they were laughing at but their laughter is what causes the most joy. A lot of people report getting a tingle from the laughter of children with their parents, followed by a drop once it stops. Their laughter was a reminder of our secluded youth and we remember it for a short time after the laughter had ended.

One of the most common questions is what is it that causes us to fade away that youthful energy that can make us smile 300 times per day? One reason is maturity paired with responsibility. Another reason is that we begin to limit our imagination.

Take a moment to imagine isolated communities of people living in areas in South America or Africa. Adults in these communities might not be as laughing often as their kids However, people who visit them describe their lives generally as joyful, lively and eager to explore the world around them.

What's the difference? One of the most striking differences is the educational system. Do not be fooled; the educational value is enormous. But there are costs of education that goes over the amount of money that is that is spent every year. The process is, by its very nature, limiting. Do this: Sit here, do this, remain within the lines, and you're not able to make use of that word 2+2 does not equal 22 and so on. We acknowledge the value for the use of arts and culture in education but that's limited. We don't advocate for an alternative system since we don't have a pre-made idea to replace it. The goal is to raise awareness and begin considering ways to unleash the best aspects of the imaginative, creative and playful distinctive you.

The objective is to restore a sense of fun to our lives. In a general sense, but with a specific intention at work. Be aware that work is a major aspect of our lives, therefore it's important to play occasionally.

Being able to play with our colleagues creates enjoyment. We will talk about this later this lets us become the Aspiring Alpha brain, where breakthrough Thinking(tm) can flourish.

The power of fun that is planned helps us to release the stress we accumulate. It stimulates our childlike thinking and allows us to invent just like kids do to come up with extraordinary solutions. It is important to be deliberate in the workplace. Most people are able to to reduce anxiety at work through an enjoyable way but this usually has negative effects. They are gossip groups and cliques which create silos and cause disconnect. Intentional playfulness and fun can create a different kind of environment. This may seem like a common sense idea in light of what we've just talked about However, once play becomes an integral part of our culture, it is easy to get away from groups that are apposed to one another into groups.

Just as an ensemble in an act. Ensemble members are aware that no matter what their job their primary goal is to make sure everyone succeeds. The reason they do that is they are aware that everyone else in the group will be focusing on their success as well.

Consider the best sports teams. The best teams of all time have one or two exceptional players among the many ordinary players. If teams have several superstars They often have a difficult time because superstar players become discontent and complain that they, individually do not get enough time, time and opportunities. However, when teams work as a team that they all win.

How do we get there? How do you make the best team , without being concerned about the top players? The answer is easy; you have people having fun and laughing.

In our next discussion When we talk about playing or having fun, we should think of it as intentional when we think about work. It doesn't mean that spontaneity is not good or should be ruled out. Actually, it's the most enjoyable type of entertainment when it is incorporated to achieve a common objective. This means that you must be thinking about having enjoyable. In light of all the things to think about, it is a good chance of having something to do.

Fun can help build bonds when there is a combination of negative and positive interactions. Particularly, you will find individuals who are doing this in five positive to one negative ratio. It doesn't mean every interaction is positive.

The Dr. John Gottman and Dr. Robert Levenson have studied the marriage of couples for years and discovered that the ratio of 5-to-1 has an incredible correlation in the study to an effective partnership. We spend lots of time with our colleagues. If we could utilize that positive-to-negative ratio, and set clearly defined priorities, the groups that we could form could be invincible.

When something isn't going well for example, a disagreement that is personal, an irritating habit and not including the other person in a major decision, or not inviting someone out for lunch and so on. It takes at minimum 3 or 4 positive actions to counteract the negative. In addition, there is a bit of extra to help the relationship remain strong. It is important to have positive feelings to withstand the storms. The connection shows how powerful negative interactions that usually trigger anxiety, as well as the secondary emotions that accompany it, could be. Even if the interactions are positive, it could cause a massive slide if something goes wrong in the future that is impossible to overcome. We must have both negative and positive interactions in harmony.

It could sound odd the idea that a proportion of interaction could result in this kind of bonding. However, imagine two kids in the playground. Every day , they play together. They draw inspiration from their imaginations and eventually become the most wonderful of friends. There is a lot of give and take with their games which creates the balance which we discussed. Then, one day, they get into huge arguments over something. They get to their homes telling their family members that they will never play with the child in the future. They talk about how awful they feel as a result of the way they were wronged by the other party. On the following day, when the children return home they'll more than likely talk about playing with their favorite friend once more.

The potential of playing with others and the chemical release it can bring to our bodies and minds is astounding. It allows us to relax and begin to think about the possibility of making changes. If we're having fun, it is more relaxed and in a good place.

Imagine someone that you do not like working with. Maybe as an administrator, you have two members of your team who are not compatible However, you require them to cooperate. What did they last had a laugh together? If the answer isn't yes you've now got the capacity to change things. It's a lot of work to achieve that however the bonds will develop as you experience five positive interactions for one negative one as a regularity.

There will be some great solo artists who will remain the only one in the room regardless of what you do? Absolutely. Provide them with awareness, assistance with guidance, laughter, and awareness. If they're an ongoing, intentional annoyance for the company that is striving to grow as one, eliminate them. The talent of a person isn't worth the continual destruction of the entire.

Being Open And Present

"When we are open to the present We receive the present as a gift for tomorrow."

Two other crucial aspects that are a part of having fun and let your creativity shine through. You are open and open. Let's look at each one of them.

Being vulnerable can be a bit terrifying. It can lead to changes and the idea of doing more of what you have to take care of which can mean giving away something you want to do. It doesn't need to be that way, and should not be. Consider trying on a new outfit. You're in the position of deciding if it's an appropriate fit is comfortable, comfortable, and looks great. If it does not do these factors for you then you get rid of it since it doesn't feel right for the cost. It's put away and you go on with no more wear.

As we travel through this guide, we'd like to ask to do this. Take a chance to try things. Take a risk and give them a fair chance without preconceived ideas. Certain aspects may be beneficial to you, while others will not. Accept what you like best However, should you find yourself stuck later on, you can come back and revisit the strategies that didn't work prior to. It will be uncomfortable.

Being uncomfortable can lead to discoveries you couldn't have thought of. When we step outside of our familiar zones, we're experiencing things differently. It's the difference between these experiences that will reveal our potential, as you'll learn during this journey.

Being present means staying in the present moment. It's about dedicating yourself to the present moment right in front of you. More importantly, what is happening right in the way. The breakthrough Thinking(tm) requires you to take the moment to remain part of the group, not in text messages, emails as well as phone calls. It's easy to get distracted and be thinking about the work we are putting in. But, staying active is a lot more enjoyable when you're having enjoyable.

If you're a manager in your organization or the one who is coordinating the group to lead the discussion, you may create some harm because you aren't fully present. You're expressing that the issue at hand isn't of any importance to you. You could be showing without intention, that opinions of others aren't worthwhile to you. One of the fastest ways to transform something everyone wants to tackle into something that people fear is when leaders don't demonstrate the importance they see in it.

If you're in the present, you are fully engaged in the events happening right before you. If you suspect that you're distracted by a call on your mobile, switch it off or put it somewhere that you won't be able to see you or hear. If you're in a meeting that's going on, but it hasn't captivated your attention, think about two questions. "Why do I attend this particular presentation?" and "What value can I add the presentation?" They will change your perception of the subject and the people around you. This is the role of leaders when they're present.

A crucial distinction needs to be made in this case. Certain of you could be a leader, regardless of title, or not, but you should be aware that if you are wearing the title of a position that suggests you might not be considered a leader. The term "leader" does not refer to an employee, supervisor or vice president and even C.E.O. These titles permit people to dictate to others what they should do. However, a leader encourages individuals to take responsibility. Leaders inspire others through presenting a vision of what's required and, if necessary, taking the time to allow others to prosper. The fastest way for people who have a title to lose respect for their position as an individual leader is not having presence.

Being an authority means that you influence a discussion with your team. The old saying "with immense power, comes with great responsibility" is valid. Your obligation is to be aware of the impact of your words, especially when your culture is new or in transition. Whatever you say, despite your intentions the words you speak could result in unintended consequences.

Another reason to be present is your customers, that we'll discuss extensively beginning in the third section. Have you ever had the experience of being with a client and they didn't seem to be engaged by the content you presented? What did you think? Stop for a second and realize that everyone you meet with is a customer since they're all humans. What you and everyone else is selling continuously are their opinions, expectations aspirations, desires and motivations. Engaging with people who surround you, regardless of whom they may be, you'll discover it easier to demonstrate your worth, regardless of who is buying.

Another aspect to being in the present moment is to maintain focus. Focus is to be aware of yourself, the people surrounding you, and those whom you are able to develop solutions for. Take a look at "The Five Letter F Word(tm)" If you want to know the ways that focus can change not just the project, but also an entire business.

Being brave is easy However, it's not easy.

"The alternative to bravery within our culture isn't cowardice, it's conformity."

Rollo May/Earl Nightengale/ Robert Anthony

As with all species that inhabits the earth the first thing we must consider is survival. In addition to our basic requirements that we have, we are social animals. We need to feel comfortable and accepted by the world as many other mammals. But, what is unique to us involves the requirement to appear precise. If you bring those three characteristics (survival being accepted, fitting in and being at peace) into perspective it will help you recognize the ways to improve your company as well as your family, team, and, perhaps you. We'll look at these aspects and explain how mastering these will result in Breakthrough Thinking(tm).

Let's think about for a second the need to be right and what others might refer to as pride or the ego. It's an extremely powerful emotion, and is it is rooted in one of the strongest emotions that is fear. The fear is the way that people view us in a negative way. If we make a mistake and we make mistakes, we could be viewed as weak. seem weak.

When people you are in contact with regularly, or at least often, fail to make a mistake are you inclined to dismiss them? Most of the time, the answer is yes. The reason you aren't is just being incorrect. It's an error. Perhaps it's just not knowing the answer. In general, we don't focus on the mistakes that others make unless they're often wrong all the times. Then why do we dwell on ourselves even when we're not?

No matter if you're a leader or believe to be one, not being right is a challenge to what we believe are the normative traits for leaders in society. The perception we make ourselves creates the desire to be the right person. When this need transforms into pride or ego, then the exact thing you're trying to avoid appears to other people.

They see you as just focused on yourself and your righteousness. This perception could undermine trust that you are able to be open to other opinions that are required for being part of the society that's not an religious cult.

There's a side to this idea. What happens when we're in a large group, and have different solutions than the other people? Think of the classic classroom scenario of being 14 years old in a math class. Everyone knows the answer that the teacher has, except for you. You double-check your math and are unable to find the error. Are you going to raise your hand and declare you're not alone and everyone else is wrong? Perhaps raise your hand to say your own opinion? Or perhaps you want the problem to be shown?

A majority of people will not do any of these things. They'd prefer to pretend that they've got the same thing as the rest of us. They are afraid of being wrong or perhaps being seen as an outlier in the crowd. They will accept the group mentality.

There are those who aren't willing to admit they're wrong. They could argue at any moment, or wait until after the class has ended to demonstrate their skills.

A small number of people would like to stand out so that they do not belong to the large group. They might say something to make themselves stand out. But they are still part of an organization You will notice.

There is also the tiniest group of people who want to learn and accept that they might be wrong.

What makes people simply join the largest group? Over time we have demonstrated our survival on the basis of the fact that we are part of a larger community. This is not a surprise because it brings many advantages to us, like sharing resources, security by numbers, readily available sexual partners so on. This social survival isn't just for humans. However, it is important to comprehend the community and social bias of the reasons we do certain actions and the reasons why they make the process of change difficult.

It is also important to recognize that being right in general doesn't matter when you come up with a fresh idea and the team that will follow it. What is important is creating an environment where everyone is able to express their opinion and not be restricted for doing so. To achieve this it is necessary to bring awareness and actively seek to alter the perceptions of each individual about what it means to be a part of the community.

You must empower the majority of people to have an alternative opinion or an opinion that isn't popular, because it is against their nature. If they are required to prove their point, you need to encourage them to check their self-confidence and demonstrate that diversity in thinking is not a matter of the right or wrong. If you want to stand out and appear unique, you must inspire them to be who they are yet contribute. For those who seek to know the "why they are here, you must ensure they are encouraged and motivated. They're often the ones who can act as the glue that helps to connect people working together since they are naturally inclined to understand the reasons why people think or feel like they do.

If you can instill this courage to everyone, it grants them the capacity to challenge concepts by seeking to comprehend. Individuals, teams or groups, as well as communities are transformed when people are comfortable overpowering the natural sensation that butterflies feel in their stomachs that result from stepping outside of their familiar zone. Maybe your community is part of your organization however, is it possible that it's bigger than that? what would be so amazing?

So, how can we encourage those people to be courageous? It is essential to show, that they are secure. The quickest way to demonstrate this is to play playing with them. When we play with each other it generates a sense of energy that shows trust, safety and a sense of comfort with one the other. The most simple way to bond is to laugh and we'll talk about this throughout the book.

The most difficult group is the ones who wish to be different in order in order to stand out. One extreme illustration of this in the case of Dr. Theodore Kaczynski, better known as the Unabomber. He says he doesn't desire to belong to society and does not care with the thoughts of society. It's still a way of trying to appear attractive to be accepted but in a completely different way.

He is said to request that he be addressed by the title doctor which is the title he earned in mathematics. This kind of dedication is typically about appearance because his brilliance makes him stand out from the rest of us. But, this isn't exclusive to him.

His distinctiveness, and the reason the world is using him as an example because there's a certain group which he wishes to be respected by. He is looking to become an example for those who are determined to sabotage the industrial and technological movements across the globe. We know he's committed to this cause since he has refused to be interviewed, with other than someone who agreed with certain of his views and also praised the manifesto. In addition during his trial he decided not to defend himself against insanity as he believed it would harm his manifesto if he did. He was concerned that if he appeared weak, it could be negative consequences for his work.

We are certainly not suggesting to bring terrorists to the group. The terrorist is used to demonstrate that even the most ardent of the world would like to be part of. If someone who is a standout is already a part of your group typically, it's due to a valid reason. You must break down their walls, so they can be accepted as part of the community.

What can you do to break down the barriers between those who are just looking to stand out? The answer is slow and one-on-one. There's no shortcut in this. It takes patience, concentration to detail, energy, and absolutely no self-importance, and most important authenticity. Honesty must shine through to everyone. Without it, nobody will take what you do seriously. They won't invest their time in whatever you're working on.

Being authentically authentic can be difficult. It requires us to step out of the comfort zone that is socially acceptable. Comfort zone refers to a kind of homeostasis. It is the body's attempt to remain in a state of regular state. In many ways, comfort is a substance. There are chemicals that get released into our systems whenever we feel relaxed which can lead to a kind of addiction. It's one of the obstacles to changing the current status of the game.

It's going to take you to be willing to make a change and actively showing that changing isn't that difficult to help others overcome this addiction. Be bold, genuine, and become an inspiration for others by taking action. If you do the things described by this publication, even when they aren't your style, you'll inspire others to do the same.

Be aware that the only way to manage is through titles or ego, it is a sign that you are a manager, not an effective leader. In Steve Jobs' words, Steve Jobs,

"Management is about convincing people to take action they aren't willing to do, whereas leadership is about inspiring people to accomplish things that they've never imagined they could."

Be brave and stand in solidarity for everyone to win. You can too.